Major Construction Works
Contractual and Financial Management

Major Construction Works
Contractual and Financial Management

Keith F. Potts

MSc FRICS MAPM
Principal Lecturer, Construction Division
University of Wolverhampton

with additional material supplied by

Brendan Patchell

ARICS
Regional Director
Bucknall Austin, Birmingham

Longman Scientific & Technical
Longman Group Limited
Longman House, Burnt Mill, Harlow
Essex CM20 2JE, England
and Associated Companies throughout the world

First published 1995

British Library Cataloguing in Publication Data
A catalogue entry for this title is available from the British Library.

ISBN 0-582-10298-7

Set by 5 in Linotron 202 10/11 Baskerville
Printed in Malaysia

Dedicated to Lesley, Ian, Gemma and Deborah

Keep interested in your career however humble; it is a real possession in the changing fortunes of time.

Max Ehrmann
1927

———————————————————————————————

Contents

Preface

The author considers himself fortunate indeed to have been involved in the contractual and financial management of five major construction projects spanning a period of 15 years. During this time experience was gained, representing either the employer or contractor, on major building and civil engineering works in the UK and overseas.

This wide range of projects included an industrial building requiring substantial earthmoving, a North Sea oil marine terminal with a large element of specialist pipework, surface works at Europe's largest coalfield and tunnelling and deep basements on an international mega-project. Standard conditions of contract on these projects included JCT 63, ICE 5th and special conditions of contract based on the international FIDIC form.

It was observed that in practice there are no artificial barriers between the professions and the construction sectors. These challenging major projects were successfully completed using teams of highly motivated professionals often formed specifically for the task in hand.

Even though the projects were very different they all required considerable management input at the various stages in the project cycle. This included carrying out feasibility studies, pre-contract estimating, raising the finance, developing the contract strategy, selecting the appropriate tendering and bid evaluation procedures, contractor's estimating and tendering, construction programming and management of the post-contract stage including cost control and claims management.

The aim of this book is to examine these key topics with particular reference to the procedures encountered on major projects. This will involve utilising the experience gained and the vast amount of information accumulated over the years.

It was whilst working for the Hong Kong Mass Transit Railway Corporation on the third stage of the vast underground project that the author, together with other senior staff, gave a series of in-house lectures to the civil engineering and quantity surveying graduates and trainees. That series of lectures formed the idea for this book, indeed some of the material appears close to its original format.

The book covers a wide range of topics including an overview of project management on major projects followed by the contractual and financial management of the pre-contract, tendering and post-contract stages of major projects. A chapter on information management precedes a section under which the relevant contractual and financial provisions on a cross-section of typical standard forms of contract are examined in detail. The book concludes with a case study examination of a successful major project.

Appropriate tutorial questions have been included at the end of each chapter which could form the basis of undergraduate or postgraduate tutorials/seminars or for continuous professional development sessions. Each chapter contains a bibliography identifying sources of information and provides a basis for further reading and research.

Keith Potts
School of Construction, Engineering and Technology
University of Wolverhampton
Wulfruna Street
Wolverhampton WV1 1SB
UK
March 1994

Acknowledgements

The author is most indebted to the many individuals and organisations who have contributed to this book particularly the co-contributor: **Brendan Patchell**, Regional Associate Director of Bucknall Austin – Quantity Surveyors and Project Managers, who wrote the bulk of three chapters: Chapter 2 'Feasibility studies', Chapter 3 'Pre-contract cost management' and Chapter 13 'Information management'.

Specific thanks also go to the following who provided significant contributions:

Professor David Jaggar, Liverpool John Moores University, for his valued support, general comments and additional input into Chapter 3 'Pre-contract cost management'.

Professor Frank Harris, University of Wolverhampton, for his contribution of the case study in Chapter 9 'Construction programming' and his continuous encouragement throughout the past 2 years.

Robert Church, University of Wolverhampton part-time M.Sc. construction management student, for his project management assignment which reviewed a contractor's cost control system; this formed the basis of much of the second half of Chapter 10 'Cost control and monitoring procedures'.

Additionally the author would also like to thank Simon Kometa (Ph.D. researcher at the University of Wolverhampton), Bob Ness (former chief estimator of Alfred McAlpine), John Gobourne (visiting lecturer and former systems manager of Alfred McAlpine) for their useful contributions.

Thanks also go to all those other researchers, students, members of the lecturing and technician staff at the University of Wolverhampton who have also made useful suggestions and comments.

Other important sources of information and aspiration have been:

- the M.Sc. Construction Management course at the Loughborough University of Technology;

- the distance learning course 'Civil Engineering Law and Contract Procedure' offered by CTA Services (Swindon) Ltd;
- the excellent 8-day residential course entitled 'Advanced Project Management' run by Professor Peter Thompson and colleagues at UMIST.

The author is particularly indebted to the RICS, the ICE, the CIOB and the University of Wolverhampton for giving permission to reproduce their past examination questions which form such a valuable source of material in the 'Tutorial questions' section.

Finally the author would like to thank Tony Shaw, Senior Lecturer at the University of Wolverhampton, for proof-reading and passing constructive comments on all the chapters – a considerable achievement in itself!

We are indebted to the following for permission to reproduce copyright material:

Building Employers Federation for 'Review of Contents' in *JCT Standard Form of Management Contract* 1987, Joint Contract Tribunal; Elsevier Science Publishers/Chapman & Hall for extracts from *ENGINEERING LAW AND THE ICE CONTRACT* 4th edition, by Abrahamson; FIDIC for extracts from pp 23–5 *Tendering Procedure* and 'Review of Contents' from *FIDIC (The Red Book)* Fourth Edition 1987; Thomas Telford Publications for 'Review of Contents' from *ICE Conditions of Contract* 6th edition. Copyright the Association of Consulting Engineers and the Federation of Civil Engineering contractors and 'Review of Contents' from *New Engineering Contract*. Copyright the Institution of Civil Engineers.

List of abbreviations

APM	Association of Project Managers
ATP	Aid and trade provision
BCIS	Building Cost Information Service of the RICS
BEC	Building Employers' Confederation
B of Q	Bills of quantities
BOT	Build–operate and transfer
CCT	Contract control total
CDC	Commonwealth Development Corporation
CESMM2	*Civil Engineering Standard Method of Measurement* (2nd edition)
CESMM3	*Civil Engineering Standard Method of Measurement* (3rd edition)
CIOB	Chartered Institute of Building
ECGD	Export Credit Guarantee Department
E(E)C	European (Economic) Community
EIB	European Investment Bank
FCEC	Federation of Civil Engineering Contractors
FIDIC	Fédération Internationale des Ingénieurs-Conseils
GFA	Gross floor area
HKMTR(C)	Hong Kong Mass Transit Railway (Corporation)
IBRD	International Bank for Reconstruction and Development
ICE	Institution of Civil Engineers
ICE 5th	*Institution of Civil Engineers Conditions of Contract* (5th Edition)
ICE 6th	*Institution of Civil Engineers Conditions of Contract* (6th Edition)
ICES	Institution of Civil Engineering Surveyors
IDA	International Development Association
JCT	Joint Contracts Tribunal
JCT 80	JCT Standard Form of Building Contract 1980 Edition
JCT 81	JCT Standard Form of Building Contract with Contractor's Design 1981
JCT 87	JCT Management Contract 1987

M&E	Mechanical and electrical
MPA	Major Projects Association
NEC	New Engineering Contract
NEDO	National Economic Development Office
NPV	Net present value
NSC/4	JCT Nominated Subcontract Form of Contract for use with JCT 80
ODA	Overseas Development Administration
OECD	Organisation for Economic Co-operation and Development
PM	Project management
PSA	Property Services Agency
RIBA	Royal Institute of British Architects
RICS	Royal Institution of Chartered Surveyors
SERC	Science and Engineering Research Council
SMM7	*Standard Method of Measurement of Building Works* (7th edition)
TML	Transmanche Link
UMIST	University of Manchester Institute of Science and Technology

Section A

Introduction

Overview and observations

Definition of major construction works

The Major Projects Association (MPA),[1] based at Templeton College Oxford, represents a wide range of institutions including owners, bankers, consultants, contractors, manufacturers, insurers, lawyers and government. This multidisciplinary and multi-industry private organisation, membership of which is available only to subscribing members, describes 'major projects' as 'any collaborative or capital project that requires knowledge, skill or resources that exceed what is readily or conventionally available to the key participants'. The association thus defines major projects more by difficulty than by size.

However, as far as this book is concerned, we shall adopt a far more general definition for 'major construction works'. This definition would include any project within the building or civil engineering sector which is normally handled by the 'major projects' division of a national contractor's organisation, rather than by the regional office. Using this definition 'major projects' was defined by one national contractor as generally any project above £20m. in value.

Relevance of study to other projects

Many of the principles of project management were first established in the defence and chemical industries and later applied within the power industry on mega civil engineering projects. However, the basic principles and procedures so established, are relevant to most construction projects. Indeed, one viewpoint is that some major projects can be considered as a series of smaller projects executed all at the same time.

Thus the project manager on a major project may experience a lifetime's problems during the pre- and post-contract stage of a single project. So by examining the contractual and financial management of major construction works our understanding of the management of other projects can be improved.

Types of projects

Major construction works are often crucial to the success of individual enterprises and to society as a whole. Such projects often embrace the building, civil and heavy engineering sectors; examples include:

- industrial factories, e.g. Toyota car plant, Burnaston, Derby
- commercial office developments, e.g. Broadgate, London
- leisure facilities, e.g. Euro Disney, Paris
- transport systems – both road and rail, including surface routes, underground systems and bridges
- power stations: coal-fired, oil and gas
- nuclear engineering projects
- defence facilities
- water distribution systems including reservoirs
- ports and airports
- coal, iron-ore and steelworks facilities
- oil and gas platforms, modules, pipelines and terminals
- refineries and petrochemical plants, etc.

Many major construction works have a history of failure with non-completion, or massive delays and/or cost overruns. The list is legend and continues to grow: the Sydney Opera House, the Humber Bridge, the Thames Barrier, the UK nuclear power plant programme, the Channel Tunnel etc.

Some major projects have cost three or four times the budget originally anticipated and taken twice as long as anticipated to build. Indeed one earlier version of the Channel Tunnel was abandoned in 1975 after millions of pounds had already been committed by the UK Labour government.

It is, however, necessary to make a distinction between project management success and project success. Anton de Wit observed[2] that the North Sea oil projects of the 1970s were in project management terms a failure – suffering from time and cost overruns. However, following the massive oil increases in 1973 and 1979 these projects were later considered a resounding success. Conversely, a project could be considered a success in project management terms and yet a failure in project terms due to extraneous circumstances outside the control of the parties.

Distinctive characteristics of major projects

Major projects, as defined by the MPA, do have their own distinctive characteristics, even when compared to large projects. Allen Sykes, a member of the MPA, has identified them as follows:[3]

- usually owned either by government or a consortium of private sector companies, hence the project has more chance of being vetoed or delayed by one of the parties;
- significant impact on the economy and environment, considerable government involvement;

- relatively few successful ones, so insufficient people with adequate experience in government or the private sector; the case study at the end of the book considers the reasons for success of one major project, the Hong Kong Mass Transit Railway;
- generally indivisible, often cannot be built in parts;
- long investigatory/approval and construction periods;
- difficulties of access, climate, terrain and large workforces, particularly on overseas projects;
- major environmental and socio-economic impact, generating public opposition;
- major impacts on markets, major strains on the contractors and suppliers;
- impose large and special risks on parties involved, particularly vulnerable to economic recession and the fallibility of long term economic forecasting;
- difficult to finance, sponsors must be able to convince financiers that the project will indeed be completed and will be able to generate adequate funds to repay debts.

Common problems

Some of the principal reasons why many major works failed to finish on time and within budget include:

- underestimating – the complexity, the logistics and sheer difficulty of executing the work; this has serious repercussions on the cost which is often underestimated on mega and defence projects, e.g. the Channel Tunnel at £10bn. is three times the initial budget and yet still the same length!
- technological advances and uncertainty – the desire to incorporate the latest innovations and technology caused particular problems on the UK power plant projects during the period 1970–80;
- late design changes – often encountered on major projects, can be a particular problem if the design is not frozen;
- correction of design errors – impossibility of performance due to an over-designed and over-zealous specification indicates that such problems do exist;
- increased safety requirements – caused particular problems on the US and the UK nuclear power plant and nuclear installations programme;
- poor industrial relations – plagued the chemical plant and North Sea oil fabrication yards in the 1970s; building work in Merseyside also affected in the 1970s;
- adverse site conditions – often encountered on piling, earthworks or tunnelling projects;
- funding availability – particularly in developing countries;
- site acquisition problems – major cause of delay on some large office development projects in London;
- quantity increases – particularly when the B of Q has been based on inadequate information;

- shortages of materials – again mainly in developing countries but steel shortages were particularly critical during the 3-day working week in the UK during the 1973 winter of discontent;
- contractor's financial difficulties – indeed some major contractors have long since ceased trading, e.g. John Howard (contractor for Humber Bridge and the massive Howard Doris concrete oil rig);
- inappropriate contract strategy – the significance of which was previously underestimated, e.g. difficulties of using the traditional approach on major hospitals subject to a multitude of late design alterations;
- inflation and interest charges – particularly in developing countries and throughout eastern Europe;
- exchange rates – significant on overseas contracts when payment for work done is made in a devalued local currency but key materials and construction equipment are purchased from the home country with staff salaries also possibly guaranteed on fixed exchange rates;
- civil unrest/political coups, etc. on overseas projects.

Morris and Hough, in their book *The Anatomy of Major Projects*,[4] analysed all the available reports of project overruns on the record. Their findings generally concur with the reasons for project delay and cost overrun as identified above.

Observations on one major project

The latest saga, in the UK, is the new British Library at St Pancras. In 1978 it was forecast that the first stage of the library would be complete in the late 1980s at a cost of £115m. In 1993 the first phase was still being built and the cost had escalated to £450m. This project, which was reported in detail in *The Daily Telegraph*,[5] was described by Mr David Mellor, the former Heritage Secretary, as 'a major league British disaster'.

It is important that we learn from this project and do not keep making the same mistakes. The reasons for the delay and cost overrun on the British Library include:

- insufficient funds from the Treasury – cost dragged on over 12 years and was subject to raging inflation;
- political interference – in 1979 the cash-strapped Conservative Government slashed the budget from £22.5m. to £9.5m. and thereafter switched funds on and off like a tap;
- five-year delay by architect in supplying vital drawings (Audit Office Report) – following introduction of stop–go policy which forced the project to be staggered;
- no strong defender for the project in government – responsibility for it divided between the Office of Arts and Libraries and the Department of the Environment, the Property Services Agency – now PSA projects – since privatised – were responsible for the building work;
- changed design – originally designed as a wonder library then whole design changed to the mini-library now being built;
- inadequate client involvement – from 1983 to 1986 the steering committee responsible for monitoring the project did not meet once!

- no chain of command within the government – more a kind of web of relationships;
- inappropriate contract strategy – management contractor appointed 12 months after project commenced to deal with more than 150 separate contracts. This was after the PSA realised that it was a bigger project than they had ever handled before. Unfortunately, the client chose to retain control over individual contractors 'to reduce the risk to the public purse', thus confusing the lines of command and introducing excessive bureaucracy, which lost time and caused endless frustration and demotivation of both staff and operatives;
- mechanical problems with 300 km of shelving, furthermore shelves now rusting as allegedly not painted correctly; major modifications also required to be made to the air-conditioning ducts in order to achieve a 'clean' environment;
- 3200 km of electrical wiring vulnerable to short-circuiting allegedly due to stripping of plastic coating during installation;
- disputes over the supply pipes to 2000 sprinkler heads prone to condensation allegedly due to insufficient slope;
- in January 1994 it was anticipated that remedial work to air-conditioning, sprinklers and wiring could hold up the project for at least 18 months;
- *The Daily Telegraph* of 17 January 1994 reported that a confidential management committee report on the British Library project considered 'that neither the building's costs nor the schedule for its occupation can be estimated and that all predictions must be based on guesswork alone'.

Some recommendations

Managing major projects is not easy, there are no quick fixes or magic solutions to the problems. However it is important to identify those critical factors which can lead to success. Allen Sykes identifies the main factors as:

- adequate investigation in the formulation period
- necessity for constant reappraisal of viability
- commitment of sufficient resources
- extra management skill required
- the experience of the project directorate

Major projects require adequate investigation in the formulative period, i.e. from the conception to the decision to construct. The project should be designed, the contract strategy identified, all government and regulatory permissions received, long-term supplies negotiated and financing arranged. This critical stage can often take between 3 and 5 years. It is at this stage that the important professional working relationships between the parties are developed.

The project will require constant reappraisal of its viability by the project directorate, i.e. the senior management team responsible for managing and directing the whole project. All the factors critical for the project's success must be constantly reviewed, i.e. economic and financial, political and regulatory, design and construction feasibility. If necessary it may be necessary for the directorate to recommend postponement or even, in exceptional circumstances, abandonment.

Furthermore, clients must be prepared to commit sufficient resources to the project, particularly at the investigatory stage. Many clients have seriously underestimated the total cost of this initial commitment or abandonment which is normally in the range from 1.5 to 5 per cent and could be as high as 15–18 per cent on major energy projects.

Clearly the key to success, on major projects, lies with the project directorate. Not only must they be multi-disciplined with design and construction experience, they must be highly objective at all times and have exceptional organisational, interpersonal and political skills. This is particularly significant in order to manage successfully a project with a large number of sponsoring participants.

The project directorate must have experienced other successful major projects of comparable size and difficulty – there being no substitute for experience; the relevance and importance of this observation should not be underestimated.

The MPA also advises that in order to make the UK construction industry more competitive, clients should move away from the traditional adversarial contractual procedures, and place a greater emphasis on the development of more imaginative competitive practices based on long-term relationships. Such a concept, which has been tried with some success with major clients with a continuing development programme, is called 'partnering' and is briefly considered in Chapter 5.

The association further advises for there to be a greater integration of design, funding, ownership, operation and maintenance with a guaranteed performance; again one such arrangement called 'BOT contracts' will be examined in Chapter 4.

Construction project management

Successful construction projects do not just happen; they require sound management of the whole construction process from inception through to completion. This approach - called 'project management' (PM) - was first defined by the CIOB in 1979 and restated in their publication *Project Management in Building*:[6] 'The overall planning, control and co-ordination of a project from inception to completion aimed at meeting a client's requirements and ensuring completion on time, within cost and to required quality standards.'

Traditionally, particularly on industrial or long-term major projects, it was the client alone who acted as the project manager. The client conceived the project based on his needs and acquired the site if not already owned. The client raised the funds for the project and used his own in-house legal staff to arrange the legal matters.

It was the client who prepared the brief, undertook the feasibility studies, designed the project, chose the procurement route and selected the contractor – often utilising the in-house project team comprising architects, civil, structural and mechanical and electrical engineers and quantity surveyors. Outside professionals would be asked to contribute to the process depending on in-house capabilities and workload.

Primarily the development of PM within the client's organisation evolved to tackle projects which had a beginning and an end. This was independent from normal business management associated with the manufacturing sector which was concerned with management of an ongoing continuous process. The distinctive characteristics of construction projects which led to the specific development of construction project management included:

- specific objectives of completion within time, within budget and to a required quality;
- requiring temporary organisations and teams often specifically assembled for the one project;
- incurring rapid expenditure;
- operating under a demanding time-scale;
- numerous collaborating parties with a high chance for conflict;
- uncertainty involving risk and requiring problem solving;
- the client receives no benefit until the project is complete.

Now, even on traditional contracts, it is becoming apparent that PM has become an essential additional element in the construction process. Amongst the reasons for the rapid growth of PM are:

- increasing size and complexity of projects;
- increasing importance of financial control;
- urgency for early completion;
- growing amount of statutory regulations;
- increased sophistication of technology;
- recognition that the typical architect's/engineer's education and experience are generally suited to design, not project management.

Significantly, the New Engineering Contract recognises this development and separates the traditional engineer's duties into four roles – one of which is the client's project manager.

Alternative strategies

When the client does not possess the necessary in-house expertise it will be necessary to appoint an independent project manager to represent the client's best interest.

Sometimes the client, often a large public authority, may wish to retain maximum control but may appoint a project manager to work alongside, to communicate with and co-ordinate the work of, the other participants. This 'non-executive project management', which is executed by a 'project co-ordinator', has been criticised by some commentators as being less than effective.

In contrast under 'executive project management' (Fig. 1.1), the client appoints a 'project manager' and delegates to him total responsibility for the project. His duties will include planning, control and direction from the inception to the completion through concept, design, procurement, construction and commissioning. Essentially this service allows a client's department to manage their business while the project manager manages their projects. Under this arrangement the project manager is responsible for appointing the other consultants and acting as the client's 'agent' handling all matters on his behalf on a day-to-day basis.

Executive PM requires that the project manager develops considerable trust with the client and keeps him informed on all matters throughout the project. The system allows the project manager to act in a dual capacity – partly as the client's agent in a confidential role. The project manager can thus assume a strong formal role necessary for the success of the project.

The role of the project manager

As long ago as 1975 the Wood report[7] recommended, on large and complex projects, that the client should appoint a project manager to act on his behalf. The report considered that the project manager should have the responsibility for the management and co-ordination between the client, designer and contractor. The report also made the observation that the project manager should have a sound appreciation of design procedures, construction methods and construction economics, but that this knowledge should be secondary to his management expertise, decision-making ability and leadership qualities.

Since the Wood Report both the CIOB and the RICS have recognised the growing importance of construction PM as a subject in its own right. The Association of Project Managers (APM)[8] was specifically formed as a professional body in order to foster and encourage the well-being of PM. It seeks to improve the standards of management theory and practice,

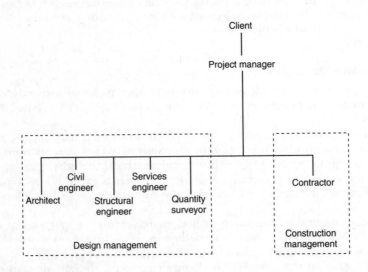

Figure 1.1   Management structure for executive project management

by direct and indirect participation in the training of its members, for example through holding regular seminars and conferences. The APM is a corporate member of the non-profit-making International Project Management Association (INTERNET) which commands considerable respect world-wide. Members of INTERNET contribute to the excellent journal *International Journal of Project Management* which is recommended reading for all those taking PM seriously.

The detailed duties of the project manager are identified in the CIOB publication *Project Management in Building* as well as in the Guidance Note published by the RICS with the *Project Management Agreement and Conditions of Engagement*.[9] In the latter document the RICS Insurance Services Limited identifies project management where the project manager appoints the other consultants, whereas in project co-ordination the client appoints the other consultants.

The project manager's principal duties include those listed below.

Initial stages with client
Establish the project manager's brief and degree of client's involvement, assessing the technical and financial viability of the project, advising on funding including grants and taxation, establishing the lines of communication between the parties and the reporting systems required, advising on site selection – arrange surveys and reports.

Feasibility stage
Co-ordinate development of the client's brief, advise on alternative designs and life cycle costing, establish quality control procedures, prepare programme from conception to completion, advise the client on the appointment of the design team and other consultants, arrange government and local authority planning approvals, establish the budget including the cash flow, identify most appropriate contract strategy, identify tendering procedures, select contractor.

Construction phase
Monitor progress, approve subcontractor selection, chair progress meetings, direct and inspect works, certify payments, check safety and quality control systems, monitor expenditure, approve variations, identify and resolve problems, issue certificates.

Completion
Ensure final accounts prepared and claims settled promptly, assist in commissioning, monitor defects, arrange for as-built drawings and manuals for client.

Table 1.1   Some observations on effective project management

Effective project management	
Is all about:	*Is not about:*
The early appointment of the PM team with clearly defined responsibilities	Appointing the project manager after the project has commenced and is already in difficulties
Ensuring that there is adequate pre-contract appraisal including identification of the risks, uncertainties and potential problem areas	Commencing the project before the pre-contract appraisal is complete
Identifying the key project objectives together with the ranking of their importance	Considering that all the objectives are equally important
Completing the design concepts before placing the orders, then freezing the design	Finalising the design after the project is let – particularly under the traditional route
Paying great attention to the selection and motivation of an adequate and experienced PM team	Appointing inexperienced staff with inadequate supervision
Choosing the appropriate contract strategy	Using the same procurement strategy as we always do
Setting up an effective communication network between the key participants before the commencement of the project	Developing a computerised communication system 12 months into the project
Learning from the past, approaching other clients with similar projects	Assuming that you have to reinvent the wheel on every project
Strict control of design change proposals vetted by review board (client's executive)	Issuing drawings without approval from executive and without identifying changes
Ensuring that the time–cost model adequately reflects the financial consequences of alternative courses of action	Considering the cost of alternatives merely on the basis of the B of Q rates
Planning and replanning, on a daily basis if critical, otherwise on a weekly and monthly basis	Pinning the construction programme to the site office wall and forgetting about it

Table 1.1 (cont)

	Effective project management	
Is all about:	*Is not about:*
Visiting the site, if necessary on a daily basis; this motivates both senior staff and operatives	Being remote and rarely visiting the site
Driving the project forward, getting in there and making it happen	Being desk-bound with a 9 till 5 mentality
Being pro-active, identifying contractor's problems early and offering solutions	Considering that the contractor's difficulties are of no concern to you
Communicating with the PM team on what needs to be done and how to do it	Being a loner and keeping all your good ideas to yourself
Thinking ahead and trying to control and influence the future	Just letting it all happen
Constant attention to detail: urging people to complete their tasks	Letting people get on with it in their own good time

Tutorial questions

1. Your company has been appointed project managers on the modernisation of Titan House, one of the largest office blocks in the City of London. The office block is fully tenanted with some 2000 people working for 47 separate firms who will remain in occupation.

    Titan House was one of London's first major reinforced concrete buildings and occupies a gross area of 42 000 m² containing 26 000 m² of office space over a garage, car showroom and banks. Unfortunately very few of the original drawings exist.

    The work comprises:
    *   complete refurbishment of all common areas and facilities on the second to eighth floor;
    *   upgrading the ground floor entrance hall;
    *   replacing all 1300 steel-framed windows;
    *   improvement of fire precautions;
    *   installation of modern sanitary facilities;
    *   new hot and cold water services and electric mains;
    *   provision for air-conditioning;
    *   a new heating system;

- replacement of boilers with new rooftop oil-fired boilers;
- upgrading the building's power supply from 1.5 to 3.5 MW;
- computerisation of lift controls (for three lifts).

The value is estimated at £40m. with the services elements accounting for 40 per cent.

   Prepare a 3000 word report on the proposed development for the owner (a joint pension fund). Make assumptions for any items not specifically identified. (Project Management Assignment, M.Sc. Construction Management, University of Wolverhampton, 1993)

2. The increasing size and complexity of modern buildings and the increasing importance of financial control and speed of procurement have led to the introduction of project management as an essential additional element of the management and co-ordination of the construction process. Discuss. (Project Management Examination, M.Sc. Construction Management, University of Wolverhampton, 1992)

3. What do you consider to be the key responsibilities of the project manager and what does each entail? (Project Management Examination, M.Sc. Construction Management, University of Wolverhampton, 1993)

4. The Chartered Institute of Building recently published its *Code of Practice for Project Management for Construction and Development*. Commentators have said that this code signals the coming of age of project management as a discipline in its own right. With this in mind answer the following question
   If there is no builder there is no building. If there is no designer there is no design. If there is no project manager what impact would that have? (Project Management Examination, M.Sc. Construction Management, University of Wolverhampton, 1993)

5. How do clients control their investments and when? (Professor P. Thompson seminar question, Advanced Project Management course, UMIST)

6. How can a contract contribute to effective project management? (Basis of a booklet to be published in 1994 by the Association of Project Managers)

7. What issues should be addressed in order to reduce conflict and litigation and increase productivity and competitiveness within the UK construction industry? (Basis of the joint government/industry review of procurement and contractual arrangements in the UK construction industry by Sir Michael Latham, 1994)

References

1. The Major Projects Association, Templeton College, Kennington, Oxford OX1 5NY
2. De Wit A 1988 Measurement of project success. *International Journal of Project Management* **6** (3), Aug 164–70
3. Sykes A 1986 Success and failure of major projects. *Civil Engineering* January/February: 17 and 19

4. Morris P W G, Hough G H 1991 *The Anatomy of Major Projects: A Study of the Reality of Project Management* John Wiley & Sons
5. Herbert S 1992 *The Daily Telegraph* 30 November, p. 4
6. CIOB 1988 *Project Management in Building* 3rd edition, p. 4
7. National Economic Development Office 1975 *The Public Client and the Construction Industries* (Wood Report)
8. The Association of Project Managers, 85 Oxford Street, High Wycombe, Bucks HP11 2DX
9. RICS 1993 *Project Management Agreement and Conditions of Engagement* 2nd edn

Further reading

Barnes M 1988 Construction project management. *International Journal of Project Management* **6** (2), May; 69–79
Barnes M 1988 Project management today, Part I, II, III and IV. *Civil Engineering Surveyor* February, March, April and May
CIOB 1992 *Code of Practice for Project Management for Construction and Development*
Hamilton B 1990 Project management. *New Builder* 21 June, 28 June, 5 July, 12 July, 19 July, 26 July
Jeffrey P 1985 Project managers and major projects. *International Journal of Project Management* **3** (4), November: 225–30
Lock D (ed) 1987 *Project Management Handbook* Gower Technical Press Ltd
NEDO 1991 *Guidelines for the Management of Major Construction Projects* HMSO, London
Thompson P 1991 The client role in project success. *International Journal of Project Management* **9** (4), May: 90–2
Wearne S H (ed) Engineering Management series, Thomas Telford Ltd
Weller P 1988 Stansted Airport: a case study. *International Journal of Project Management* **6** (3), August: 133–9
Wideman R M 1989 Successful project control and execution. *International Journal of Project Management* **7** (2), May: 109–13
Youker R 1989 Managing the project cycle for time, cost and quality: lessons from the World Bank experience. *International Journal of Project Management* **7** (1), February: 52–7

Section B

Management of the pre-contract stage

2

Feasibility studies

Introduction

One of the first duties of the project manager/engineer is to prepare a preliminary study for the project. The aim of the study, which should be inexpensive and quick to prepare, will be to examine the viability of the project. The study should involve an examination of the broad effects of the scheme, including economic justification, finance and grants and the effect on the local environs.

The preliminary study will be followed by the preparation of an in-depth feasibility study. The purpose of this study will be to produce sufficient information to enable the promoter to take the decision to proceed with the scheme. The feasibility study will involve:

- identifying the basic aims of the project;
- preparing the outline designs of various alternative options;
- identifying the initial capital and operational costs of the various schemes;
- preparing an outline programme covering the design, construction and commissioning phases including any period for public consultations;
- preparing a budget for the project;
- identifying sources of funding, grants and taxation provisions;
- recommending an appropriate course of action.

It is essential that a full economic evaluation and financial feasibility study is undertaken for any major construction work. Even if a client is not concerned with profit from the development, he will inevitably be working within general economic and planning constraints and within a tight financial budget.

The economic evaluation will involve comparing different schemes and alternatives in order to identify the most appropriate scheme. Until recently, in the public sector, it was often the social and sometimes even the political benefits which justified expenditure. Now with the UK government's increased emphasis on privatisation the justification for projects will be determined more by financial feasibility and cost–benefit analysis calculations.

At the end of the feasibility stage the project manager/engineer should decide whether or not to proceed with the project.

Scenario for a large commercial development

A typical scenario for a large commercial development comprising a speculative shops and offices complex is described below, including the parties typically involved in the scheme.

Landowner

The freehold owner normally believes that the land he owns can be put to better use than at present and is prepared to sell it.

Developer

A speculative developer will be looking for land that is under-utilised, and therefore undervalued. He will decide that the land can be developed, that is, increased in value. Quite often the developer will apply to the local authority for a change of use.

Local authority

The local authority controls the use of land under its control. It has a duty to ensure that the development conforms with its structure plan. Retail shopping centres, industrial factories and residential areas are kept separate with the 'green belt' protected. However, there is pressure for more buildings so authorities are obliged to consent to change of use where a scheme contributes to the local economy and conforms to the structure plan.

The local authority can demand that the developer contribute to the cost of feeder roads or roundabouts; these are referred to as 'Section 106 Works'. These works can be quite extensive and go beyond costs incurred directly attributable to the development. The local authority can benefit from this external income as a means to improve the environment on the back of the development.

Letting agents

The development will be of no use without the end-users. Letting agents will be employed to inform the developer of the likely rental levels attainable and the specification demanded for that rental. The rental and specification are at the mercy of market forces. The letting agents are responsible for marketing the development and receiving rent.

Design team

The design team actually designs the development from the developer's and letting agents' brief. A typical design team comprises the architect, quantity surveyor, structural engineer and services engineer. On larger, more complex developments there may be as many as 30 design team members, including highway engineers, landscape architects, rights of light surveyors, acoustic engineers, mechanical handling engineers and interior designers.

Project manager

On more modest projects the architect also performs the task of the client's project manager, but on larger projects the function is a separate one. Often project managers are imposed on the design team by the client to ensure good communication between the parties. It is essential that the project manager has the power to demand action by all the parties – the client's own departments can often be the worst offenders.

It is worth reiterating that the term 'project manager' should be applied only where the client's representative appoints the other consultants; if the client selects the consultants the client's representative is known as the 'project co-ordinator'.

Tenant

It is preferable if the developer can pre-let the building to a tenant before the work commences. This is advantageous to both parties; the developer reduces his risk and the tenant can influence the design provided he is 'on board' before the design is too advanced. Often the tenant will request that fitting-out work is included in the main contract. Initially the tenant will pay rental to the developer.

Contractor

The contractor will normally undertake to construct the works within a prescribed period for an agreed tendered lump sum. Any tenants' extras will normally be treated as variation orders and will require an adjustment to the tendered sum.

Funding institutions

Developers seldom hold on to the buildings they create. Often the development is sold on to a funding institution such as an insurance company or pension fund. The developer makes a one-off profit on the sale of the completed building. The funding institution makes a small annual profit on the rental and receives a long-term steady growth on the value of the development.

Often the parties involved with a development perform more than one of the above functions, such as owner/occupier or developer/contractor.

Financial appraisal

The development appraisal example that follows (Table 2.1) explains some of the techniques used in financial appraisal. Obviously each project has to be individually appraised but the basic format is valid for any project. More sophisticated development appraisals can be produced based on cash flow principles.

Development value

The lettable area is the area of the building that can be rented to a tenant. It excludes the areas of lobbies, plant rooms and circulation areas. Different areas will command different rents, e.g. a prime ground floor may be rented to retail outlets and cheaper upper floors for offices. The different areas must be kept separate and measured in imperial square feet as these are quoted by the letting agents.

The lettable area times the square foot rental produces the annual rent received. This annual rent is multiplied by the years purchase to calculate the equivalent value of the rental income per annum as the gross development value. The years purchase (YP) is calculated based on 100 divided by a capitalisation percentage. For example, with a capitalisation percentage of 10 per cent the YP factor is 100/10, i.e. 10, likewise with a capitalisation factor of 8 per cent the YP factor is 100/8, i.e. 12.5. Letting agents' fees and legal costs are then deducted to give the net development value.

Development costs: site

The area of land is normally quoted in acres. Sometimes there is confusion between the total land area and the developable land area. A farm may be purchased for development with say a marsh or protected woodland included in the sale which cannot be built upon, thus affecting the efficiency of the site.

In addition to the cost of the land itself must be added stamp duty, legal costs incurred in purchasing and charges in connection with building regulations and planning applications.

Development costs: construction

[handwritten note: cost of fitting out added not this]

The construction costs normally come straight from the cost plan. Often there are preparatory works carried out by the developer such as demolition, road works and site stabilisation which have to be added. The cost of fitting out for tenants is also normally included in the development costs, although in practice it is often considered an extra to the 'contract sum'.

Tenants' extras can be a major problem. They are normally excluded in the first viability calculation but once a tenant is interested can grow to a substantial sum. The cost is normally handled in one of three ways:

- absorbed by the developer as a 'discount' to entice a tenant, resulting in reduced profitability but increased saleability;

Table 2.1 Development appraisal statement

SUMMARY
CLIENT
PROJECT
REPORT 4

25TH DECEMBER 1992

DEVELOPMENT APPRAISAL

CAPITAL VALUATION *

		2: Shops	*3: Pre-let offices*	*4: Spec. offices*	*5: Spec. tower*	*Total or avg*
Car spaces		2 000	500	336	160	2 996
Lettable area	m²	29 986	34 318	14 764	5 830	84 898
Lettable area	SF	325 738	372 796	160 381	63 331	922 247
Rent (lettable) £/SF/year		**12.34**	**17.00**	**15.00**	**16.00**	**14.94**
Annual area rental income		4 019 606	6 337 539	2 405 720	1 013 301	13 776 166
Car space rental rate		0	950	950	950	2 850
Car space rental income		0	475 000	319 200	152 000	946 200
Annual rental	£	4 019 606	6 812 539	2 724 920	1 165 301	14 722 366
Capitalisation	%	7.50	7.50	7.50	7.50	7.50
Gross development value	£	53 594 745	90 833 858	36 332 266	15 537 342	196 298 212
Letting agents' fees	2.00%	1 071 895	1 816 677	726 645	310 747	3 925 964
Legal	1.00%	535 947	908 339	363 323	155 373	1 962 982
Net development value	£	51 986 903	88 108 842	35 242 298	15 071 222	190 409 266

Table 2.1 (cont)

SUMMARY
CLIENT
PROJECT
REPORT 4

25TH DECEMBER 1992

DEVELOPMENT APPRAISAL

DEVELOPMENT COSTS		2: Shops	3: Pre-let offices	4: Spec. offices	5: Spec. tower	Total or avg
Site value, costs and fees						
Site area	45 977 m²	17 011	17 931	7 816	3 218	45 977
Proportion of total	%	37	39	17	7	100
Cost/m²	£/m²	308.85	308.85	308.85	308.85	1 235
Site value	£	5 254 000	5 538 000	2 414 000	994 000	14 200 000
Enabling works	£/m	370 000	390 000	170 000	70 000	1 000 000
Land cost + enabling		5 624 000	5 928 000	2 584 000	1 064 000	15 200 000
Legal costs	2.00%	112 480	118 560	51 680	21 280	304 000
Total site costs & fees		5 736 480	6 046 560	2 635 680	1 085 280	15 504 000
Construction costs and fees						
Demolition/enabling works	m²					0
Gross floor area		37 052	38 646	16 779	6 776	99 253
Cost/m² building		551.03	1 204.39	1 023.34	1 080.41	921.42
Building cost	20 416 880	46 544 851	17 170 614	7 320 859	91 453 204	

Table 2.1 (cont)

SUMMARY
CLIENT
PROJECT
REPORT 4

25TH DECEMBER 1992

DEVELOPMENT APPRAISAL

DEVELOPMENT COSTS

		2: Shops	3: Pre-let offices	4: Spec. offices	5: Spec. tower	Total or avg
% of 1: Car park (spaces)	33.65 £m.	[67]	[17]	[11]	[5]	[100]
% of 6: Siteworks (m² GFA)	3.50 £m.	[37]	[39]	[17]	[7]	[100]
Infrastructure contribution	37.15 £m.	23 842 216	7 087 104	4 297 213	1 927 795	37 154 329
Total cost plan figure		44 259 096	53 631 955	21 467 827	9 248 654	128 607 533
Professional fees	11.00%	4 868 501	5 899 515	2 361 461	1 017 352	14 146 829
Building regulations, etc.	2.00%	885 182	1 072 639	429 357	184 973	2 572 151
Construction and fees		50 012 779	60 604 109	24 258 645	10 450 979	145 326 513
VAT	0.00%	0	0	0	0	0
Total construction costs		50 012 779	60 604 109	24 258 645	10 450 979	145 326 513

Financing costs

	%	Int %	Wks *				
	100	8.00	139	Direct payment			
Site				1 436 382	626 115	257 812	0	2 320 309

Table 2.1 (cont)

SUMMARY
CLIENT
PROJECT
REPORT 4

25TH DECEMBER 1992

DEVELOPMENT APPRAISAL

DEVELOPMENT COSTS

		2: Shops	3: Pre-let offices	4: Spec. offices	5: Spec. tower	Total or avg
Construction		50 8.00 104 Direct payment	5 238 860	2 097 013	903 424	8 239 297
Total financing costs		0	6 675 242	2 723 128	1 161 236	10 559 606
Miscellaneous costs						
Promotion/marketing		100 000	50 000	50 000	20 000	220 000
Other						0
Total miscellaneous costs	£	100 000	50 000	50 000	20 000	220 000
Total development costs	£	55 849 259	73 375 911	29 667 453	12 717 495	171 610 119
Net development value	£	51 986 903	88 108 842	35 242 298	15 071 222	190 409 266
Minus total development costs	£	55 849 259	73 375 911	29 667 453	12 717 495	171 610 119
Gross residual profit	£	(3 862 356)	14 732 931	5 574 845	2 353 726	18 799 147
	%	(6.92)	20.08	18.79	18.51	10.95

*Assumes no financing costs. Direct stage payments and sale price of 52 000 000.

- allowed for by an increase in the rental;
- to be paid for by the tenant as a capital payment in which case the development value must be increased accordingly.

Professional fees must be added to the construction cost, normally as a percentage. It is important to allow for non-construction related fees such as 'artist's impressions' and soil investigation costs which would normally be expended before construction costs.

Development costs: finance

Finance costs need to be added separately to the land costs, construction fees and professional fees. The finance for the land will normally be over a longer period and may well be at a lower preferential interest rate.

The financial cycle commences with the purchase of the land, followed by the period for design, then the actual construction period, then the void after construction – all before the building is fully tenanted. Only when the building is tenanted will the building become a profit-making commodity.

Development costs: miscellaneous

Miscellaneous costs are dependent on the type of project – unlike the other costs which occur on all projects. Marketing or promotion costs can be a considerable expenditure. Normally this entails a sign board and advertising but could include fitting-out a show office. Miscellaneous costs can also include the negative costs of grants or subsidies (see Chapter 4 for sources of grants).

Developer's profit

The total gross development minus the development cost equals the profit. It is not easy to fully appreciate the significance of a lump sum, it is therefore divided by the development cost to produce the income return on costs; 15 per cent is considered normal.

The gross capital value per square foot is the gross development area divided by the lettable area and it can be compared with the annual rental. Developers still use the old imperial unit when expressing the area of property.

Building efficiency

Site coverage

The more building that can be squeezed into a site the better from the developer's point of view. Taken to an extreme this would result in overcrowding as can be witnessed in the Victorian areas of inner cities. Nowadays local authorities insist that developments do not exceed the predefined plot ratios. The plot ratio is calculated as the area of the building on all floors divided by the area of the site; from the developer's viewpoint the higher the figure the better.

A three-storey office building with a building footprint of a third of the area of the site would be 100 per cent. A ten-storey city tower consuming the entire site would be 1000 per cent. An out-of-town single-storey retail development on a quarter of the site would be 25 per cent.

Typical plot ratios defined by local authorities are between 25 per cent and 40 per cent. The area used for this calculation is the gross building area measured to the external face of external walls including the circulation area. This is quite different from the lettable area which is the net building area measured to the internal face of external walls excluding circulation areas.

In the example (Fig. 2.1) the 'Nat West' tower in the city area is ten storeys high and consumes 100 per cent of the site area. It therefore has a plot ratio of 10 – a very high figure and is obviously an efficient use of the site. The high cost of land in city centres demands greater utilisation of the site otherwise the development equation does not prove viable.

Lettable/gross ratio

The lettable/gross area ratio (fig. 2.2) is the lettable or rental area divided by the gross area. A figure between 70 and 80 per cent is normal; from the developer's viewpoint the nearer 100 per cent the better.

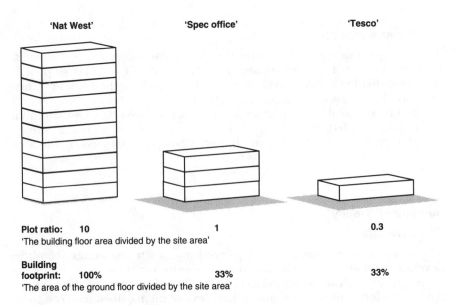

| 'Nat West' | 'Spec office' | 'Tesco' |

Plot ratio: 10   1   0.3
'The building floor area divided by the site area'

Building
footprint: 100%   33%   33%
'The area of the ground floor divided by the site area'

Figure 2.1   Site coverage ratios

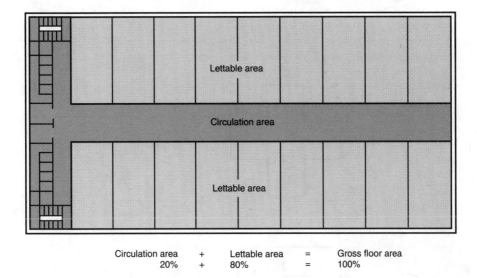

Circulation area   +   Lettable area   =   Gross floor area
20%   +   80%   =   100%

Figure 2.2   Lettable/gross ratio

In an owner/occupier building used as a headquarters building the lettable area may be as low as 55 per cent. This example may be extremely environmentally attractive but financially a disaster, particularly as it would be virtually impossible to sell to a funding institution.

Plan shape

Different plan shapes result in different efficiencies of enclosed space. The relative efficiencies of different plan shapes is measured by the wall/floor ratio, i.e. the length of the wall divided by the area of the floor enclosed. Theoretically the best shape is a circular building, but in practice it would be difficult to build and therefore expensive. A much better shape is the square, the more complex the plan shape the less efficient the building envelope (Fig. 2.3).

However, this whole issue of the most economic plan shape has been somewhat turned around by the growing trend away from air-conditioned developments to more environmentally friendly buildings which rely on natural ventilation. Indeed, plan shape 'A' may prove to be the most expensive option when mechanical and electrical services are considered.

Cost–benefit analysis

The cost benefit for the speculative office development described above is based on a financial equation which reflects the capitalist free market forces of supply and demand. If the tenant is prepared to pay sufficient rental to give the developer a profit the developer will construct the building.

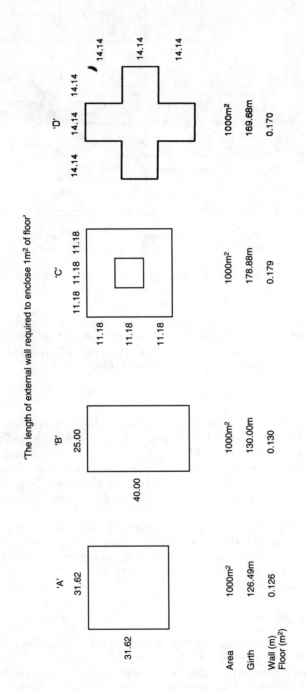

Figure 2.3   Efficiency of plan shape

Cost–benefit analysis can be considered another tool in the decision-making process and is used mainly in the public sector. An attempt is made to calculate and compare total costs with the benefits which are likely to accrue – many of which are intangible and difficult to evaluate in financial terms.

As an example consider the cost–benefit analysis of a proposed major urban dual carriageway given below.

Costs

Total costs including all compulsory purchase, construction, all professional fees and other related costs are estimated at £75 million.

Benefits

The scheme will benefit motorists by virtue of fuel, time, accident savings and reduction in driver stress (based on a 30-year period at 1994 prices discounted at 8 per cent (Table 2.2). The value of accident savings, which includes traffic accidents to pedestrians and cyclists, may be better expressed as a reduction in the number of casualties, over a 30-year period (Table 2.3).

Table 2.2   £75m. urban road improvement scheme – potential economic savings

	High economic growth (£m.)	Low economic growth (£m.)
Car users	90	67
Users of light goods vehicles	18	13
Users of other goods vehicles	7	5
Bus operators and passengers	5	5
Value of accident savings:		
All vehicle users	34	25
	154	115

Table 2.3   £75m. urban road improvement scheme – potential accident savings

	High economic growth	Low economic growth
Fatal	60	50
Serious	1250	950
Slight	4500	4250

Obviously the main problem is in identifying a realistic cost saving to set against the reduction in fatalities and accidents. Allowing £1.75m. per fatality, as the government's guidelines, the scheme is clearly justified on a cost–benefit basis whether economic growth is high or low.

Environmental impact study

In a civilised society there should be a system of checks to ensure that a proposed development does not have an adverse effect on a community. Totally free market forces should not be allowed to decide whether or not a development takes place. National parks, the green belt and listed buildings are all protected from excesses of development.

The EEC Directive on environmental assessments (85/337/EEC), enabled by the Town and Country Planning (Assessment of Environmental Effects) Regulations 1988 and the Highways (Assessment of Environmental Effects) Regulations 1988, requires all the environmental effects of major schemes to be assessed and described. Such environmental effects should be taken into account at the earliest possible stage in all the technical planning and decision-making processes.

There are nearly 60 types of development where an environmental impact analysis must be carried out if there is likely to be a significant, adverse or beneficial, effect on the environment. The assessment should cover all significant impacts of the scheme on human beings, fauna, flora, soil, water, air climate, material assets, cultural heritage, landscape and the interrelationship of these factors.

For example, on Sizewell C pressurised water reactor nuclear power station in Suffolk an environmental statement was produced by Nuclear Electric. The environmental statement assisted the Secretary of State and the local authorities in formulating their views on the consent application and informed the members of public on matters in which they might be interested or about which they might be concerned. The non-technical environmental statement for the Sizewell C project contained the following sections:

1. Introduction and background to the proposal to build Sizewell C:
   (a) background
   (b) consent procedure
   (c) the environmental statement
   (d) choice of Sizewell C
2. Project description:
   (a) site location
   (b) the pressurised water reactor
   (c) construction activities
3. Environmental effects:
   (a) effects on land use
   (b) effects on flora and fauna
   (c) effects on marine life
   (d) architecture and landscape
   (e) socio-economic effects

(i)   employment effects
(ii)   accommodation effects
(iii)   effects on the local services
(iv)   other economic effects
(f) transport effects
(g) effects of noise
(h) safety
4. Other environmental aspects:
(a) air quality
(b) climate
(c) archaeology
(d) coastal geomorphology
(e) methane
(f) other aspects

Such non-technical environmental statements are frequently produced by promoters of public works or where there is a public interest.

Contaminated land

As the land supply is finite many developments inevitably will involve construction on contaminated sites. Contaminated land is land which has been polluted by industry, urbanisation, or past development in a way which makes it impossible to use without treatment.

In 1992 the UK government attempted to introduce a national system of registration for contaminated land under the Draft Environmental Protection Act 1990 (Section 143 registers). Unfortunately the implementation of such a register has been delayed by the property and financial institutions primarily concerned with the effect on property values. However, the ensuing debate has meant that the perceptions about contaminated land in the UK have been radically altered. Developers will in the future take a far more cautious approach to the acquisition, redevelopment and management of sites containing contaminated land.

Risk analysis and management

In the past often little attention was paid to the identification and management of project risks. It was therefore not surprising that many major complex projects were completed late with massive cost overruns.

In construction projects each of the three targets of time, cost and performance quality are subject to risk and uncertainty. Sound PM should now ensure that the risks and their effects are considered at all the key decision points throughout the project.

Risk management, which is far from an exact science, should ensure that only those projects which are economically viable or worthwhile will be approved. The process involves:

1. *Identifying the risk factors likely to occur on the project*, e.g. unforeseen ground conditions, effects of inflation, variations in interest rates, likelihood of change, default of contractor or subcontractors.

2. *Analysing the risks*, i.e. assessing the probability and extent of each of the main risks (between six and ten in total). The analysis can take place in a forum involving all the key project participants. Events that have a remote chance of occurring are normally excluded. Once the range of probabilities for each of the main risks has been identified it will be necessary to combine the results in a way that reflects the overall probability. Many of the larger multidisciplinary clients in the private sector analyse projects using computer software utilising one or more of the many methods of calculation including: sensitivity analysis, probability analysis, Monte Carlo simulation, decision tree analysis or the utility theory. When completed the simulation should be presented in an understandable form from which the client and the project management team can make decisions.
3. *Deciding on a response for a risk or a combination of risks.* This may involve avoiding, or reducing, transferring or retaining the risk(s). As a general rule if a risk is usually severe and outside the control of the contractor then it should be carried by the client.

Risk management, whilst being most important at the viability stage, should continue through the life of the project. It should enable the project team to decide the brief after evaluating the various schemes at the appraisal stage, preparing the final proposal for funding and deciding the contract strategy.

Practitioners who are new to risk management will gain considerable benefit from identifying the most likely primary source of risk, identifying the options and assessing the advantages and disadvantages and implications before, for example, recommending the appropriate contract strategy.

Life cycle costing

The life cycle cost of an asset is defined as the total cost over its operating life, including initial acquisition and subsequent running expenditure. This approach enables the client's adviser to:

- identify the total cost commitment
- make a choice between alternatives
- consider the operating costs of the assets
- identify those areas where costs may be reduced

The technique which involves 'net present value' calculations can be applied to new or to existing buildings or structures.

Swanston described the relevance of life cycle costing when comparing the total cost of ownership of an existing elderly people's home.[1] The technique helped persuade Cheshire County Council to invest in replacement property. However, Ashworth[2] identified that the application of life cycle costing in practice may be somewhat limited due to the uncertainty of forecasting the future; for example, who could have forecast that the high-rise residential tower blocks built in the 1960s would be demolished within 30 years!

Cash flow

Major construction projects are built, and paid for, over a long period. The timing of the actual payments made by the client to the parties is often critical. Clients seek to balance their books within their standard accountancy period. Whenever possible negative construction costs should match positive income. The essence of a cash flow is to ensure that a company is not exposed to excessive negative cash flow.

In practice, design teams and contractors are often put under considerable pressure to commence work on site before Christmas and complete within unrealistic contract periods. Very often the client has to spend a fixed amount of his capital budget within a financial year, rarely will he be able to take an unspent budget to the following year.

'S' curves

The simplest way to produce a cash flow is by using the 'S' curve method on a computer spreadsheet. Expenditure on construction works often follows the shape of a 'lazy S' curve reflecting a slow start, hectic middle and slow finish (Fig. 2.4).

It is possible to analyse cash flows by comparing them in percentage terms. A chart can be created showing percentages of construction period plotted against percentages of construction cost. It may be useful to plot the cash flow from several previous similar projects in order to identify the typical 'S' curve for the type of project in question.

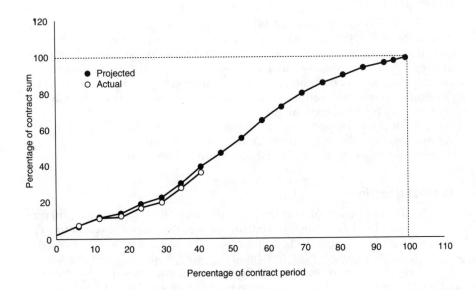

Figure 2.4 'S' curve cash flow

The Department of Health has carried out research into payments based on an 'S' curves on health buildings. Likewise the PSA analysed 500 projects before including the standard 'S' curve in the GC/Works/1 (Edition 3) conditions of contract. The great advantage of payments based on 'S' curves is the speed with which the payment due to the contractor can be identified. The difficulty comes when the project deviates from the anticipated programme, e.g. due to unforeseen ground conditions or late issue of drawings.

Bar chart cash flows

A bar chart cash flow is more sophisticated but takes more time to produce. A more accurate cash flow can be produced based on a priced contractor's programme for the project. This method has the advantage that the cash flow will more accurately reflect the different types of building and different forms of construction which can severely distort the classic 'S' curve.

The bill of quantities (B of Q) is broken down in cost in the same format as the contractor's tender bar chart programme. The cost for each activity is then divided by the number of weeks for that activity and the costs allocated to the appropriate weeks on the overall contract period. The product of the individual weekly subtotals is then calculated to produce the total cost for each week.

This type of cash flow can be adjusted to match actual performance on site and also take into account exceptional cost items that would not be reflected in a simple 'S' curve.

Developer's cash flow

The developer has to forecast all cash flows, not just the contract sum. The costs are first subdivided into positive and negative flows. Positive cash flow, or income, will include land grants, subsidies, rental and sale to funding institutions. Negative cash flow, or expenditure, included land purchase, construction costs, fees and finance charges. It is critical to the developer to not only reduce expenditure but to ensure that his cash flow conforms to his budgeted financial plan.

Many developers and finance departments work in 13 equal 4-week accounting periods rather than months. It is essential that the cash flow information matches the financial discipline of the client.

Tutorial questions

1. On the boundary of the leisure park, next to a motorway junction, freehold land is being offered for sale. A developer intends to erect an office block of 8000 m². Planning permission has been obtained.

   It is anticipated that the building will produce a net income of £800 000 p.a. and will cost £800m. to build. An allowance for external works of £100 000 needs to be included. The yield for the office space of this kind is to be taken as 5.5 per cent.

It is assumed that building work will take 18 months and that work on the site could start 6 months after the purchase of the site.

(a) Making realistic assumptions about any information not provided above, prepare a residual valuation of the site on behalf of the developer.

(b) Explain the term 'net lettable floor area' and describe the effect it has on the above valuation. Outline ways in which design can affect the efficiency ratio for projects of this type. (RICS Final Examination (Quantity Surveying), 1990)

2. The use of cost–benefit analysis (CBA) has been criticised as being simply a means of justifying uneconomic public sector expenditure. To what extent might it be used in the private sector to support the developers in negotiations with a local authority either over contentious planning decisions, or issues related to planning gain concerning the marina development? (Please make all necessary assumptions). (RICS Final Examination (Quantity Surveying), 1991)

3. You have been asked to contribute to an impact study covering both the social and economic implications of a proposed leisure park development for the community and surrounding area. Draft an outline of the report covering what you consider to be the main issues. (RICS Final Examination (Quantity Surveying), 1990)

4. Discuss the potential for the application of risk analysis techniques in relation to speculative development appraisal. (RICS Final (Quantity Surveying), 1987 syllabus, specimen paper)

5. With the redevelopment of landfill areas there are elements of risk: the treatment of hazardous or toxic matter which may have been dumped illegally, and the longer-term consequences of methane, carbon dioxide and leachate emissions and the potential for ground settlement

As the public perceptions of these risks is likely to have a considerable impact on property and rental values, potential investors and users might be discouraged. Discuss measures which might be used to allay such fears and restore public confidence. (RICS Final Examination (Quantity Surveying), 1991)

References

1. Swanston R 1983 New or old in Cheshire. *Chartered Quantity Surveyor* August
2. Ashworth A 1988 Making life cycle costing work. *Chartered Quantity Surveyor* April

Further reading

Ashworth A 1988 *Cost Studies in Buildings* Longman Scientific & Technical
Association of Project Managers 1992 *Project Risk Analysis and Management – a Guide* APM
Flanagan R, Norman G 1989 *Life Cycle Costing for Construction* Surveyors Publications, RICS

Flanagan R et al 1989 *Life Cycle Costing Theory and Practice* BSP Professional Books

Gruneberg S, Weight D 1990 *Feasibility Studies in Construction* Mitchell

Institute of Highways and Transportation with the Department of Transport 1987 *Roads and Traffic in Urban Areas* HMSO

Institution of Civil Engineers 1976 *An Introduction to Engineering Economics* ICE

Perry J G, Hayes R W 1985 Risk and its management in construction projects. *Proc Instn Civ Engrs* Part 1, **78** (June): 499–521

Perry J G, Hayes R W 1986 Risk and its management in construction projects. *Proc Instn Civ Engrs* Part 1, **80:** 757–64

Seeley I 1983 *Building Economics* 3rd edn) Macmillan

Thompson P, Perry J 1992 *Engineering Construction Risks* Thomas Telford

Yates A 1986 Assessing uncertainty. *Chartered Quantity Surveyor* November: 27

3

Pre-contract cost management

Introduction

In order to provide effective pre-contract cost management there are two essential components:

1. The establishment of realistic budgets through cost estimating;
2. Ensuring compliance with budgets as the design evolves through the process of cost control.

Construction is a major capital expenditure which clients do not commence until they are certain that there is a benefit. This benefit may be for society in the case of public projects or purely based on financial considerations in the case of private projects.

Most clients are working within tight predefined budgets which are often part of a larger overall scheme. If the budget is exceeded or the quality not met then the scheme could fail. Pre-contract estimating sets the original budget – forecasting the likely expenditure to the client. This budget should be used positively to ensure that the design stays within the scope of the original scheme.

Pre-contract cost control in the building sector

During the 1950s as a means of controlling costs and establishing greater reliability of pre-tendering budget estimating, the technique of design cost control based on elemental cost planning was established. This technique is now well established and has been further developed by the Building Cost Information Service of the RICS (BCIS) to include a national data base of elemental cost analyses which can be accessed using on-line computer techniques. Such information can be used to aid the pre-contract estimating process in the building sector as well as helping to ensure value for money by aiding the designer to ensure the most appropriate distribution of costs within the project.

The basis of design cost control using the above technique is the analysis of existing projects into functional elements in order to provide a means of comparison between projects either existing or planned. A building element is defined as a part of a building performing a function regardless of its specification. For example, the element 'frame' could be structural steel, *in situ* or precast concrete, or timber. Even if a material is selected, such as concrete, various techniques may be suggested such as wide spans, waffle floors, etc. Whichever solution is eventually chosen its function must be performed within the budget set aside for it. Thus elemental analysis allows you to compare the costs of the same elements between two or more buildings.

There are two common elemental systems used in the UK; the CI/SfB which was adopted by the RIBA from the international SfB system which was first developed in Sweden in the late 1940s and the BCIS which was developed by the RICS and is more commonly used by quantity surveyors (Fig. 3.1).

As the cost element under consideration is performing the same function then an objective assessment can be made as to why there may be differences in costs between the same element in different buildings. There are three main reasons why differences in cost occur:

1. Differences in time: for example, the same building built in 1993 will cost more than if it had been built in 1983; such differences can be accommodated by the use of indices.
2. Quantitative differences: one building may have more of the same element than another, e.g. twice the area of external walling.
3. Qualitative differences: one building has hardwood internal doors and another softwood internal doors.

The above adjustments are further described in this chapter under the section on cost analysis.

Elements subdivide the costs for a single building. Therefore on major projects it is necessary to consider individual buildings or parts of buildings. A major shopping centre may be split into common basement, finished malls, unfinished shells, hotel and car parking. The parts of the whole may be physically linked and difficult to separate, but separation will ease estimating and control. The costs of the identifiable parts can then be compared against other schemes. A composite rate per square metre is meaningless when you mix the cost of finished atrium malls with unfinished shells.

As well as separating out parts of the building that serve different functions it is important to separate for phasing. Many major projects have to be built around existing structures which increases the cost because of temporary works as well as inflation.

Budget estimating techniques

There are four major ways to estimate the cost of a building during the design stage which are dependent on the quantity and quality of information available at the time the estimate is required. Figure 3.2 illustrates when each technique can be applied, how accurate they are and how long they take to carry out. As a general principle, estimates

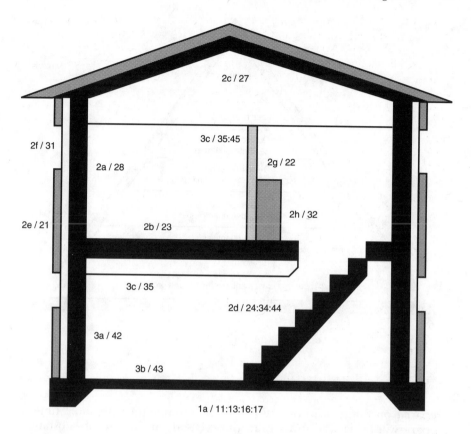

BCIS				CI/SfB			
1a	Substructure	5b	Equipment	11	Excavation	44	Stair finishes
2	Superstructure	5c	Disposal	13	Floor beds	45	Ceiling finishes
2a	Frame	5d	Water	16	Foundations	47	Roof finishes
2b	Upper floors	5e	Heat source	17	Piling	51	Refuse disposal
2c	Roof	5f	Heating and air	21	External walls	53	Internal drainage
2d	Stairs		treatment	22	Internal walls	54	Gas
2e	External walls	5g	Ventilating	23	Floors	55	Refrigeration
2f	Windows and	5h	Electrical	24	Stairs	56	Space heating
	exterior doors	5i	Gas	27	Roof	57	Ventilation
2g	Internal walls	5j	Lift	28	Frames	61	Power source
	and partitions	5k	Protective	31	External wall	62	Power supplies
2h	Internal doors	5l	Communications		openings	63	Lighting
3	Internal finishes	5m	Special	32	Internal wall	64	Communication
3a	Wall	5n	BWIC		openings	66	Transport
3b	Floor	5o	Builder's profit	33	Floor openings	68	Security
3c	Ceiling	6a	Siteworks	34	Balustrading	72	General fittings
4a	Fittings and	6b	Drainage	35	Suspended ceilings	74	Sanitary fittings
	furnishings	6c	External services	37	Roof lights	91	General siteworks
5	Services	6d	Minor building	41	External wall	92	Minor structures
5a	Sanitary		works		finishes	93	Enclosures
				42	Internal wall	94	Surface treatment
					finishes	95	Drainage
				43	Floor finishes	96	Installations

Figure 3.1   BCIS and CI/SfB elements

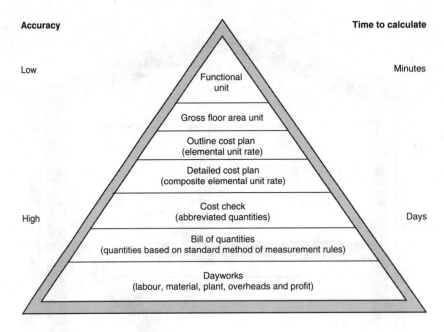

Figure 3.2   Estimating techniques triangle

prepared on minimal information and those taking a very short time to prepare will be less accurate than those based on the use of substantial information and needing considerable time to prepare.

The techniques in common use are:

1.   Function or performance-related approaches;
2.   Size-related approaches;
3.   Approaches based on the manipulation of elemental cost analysis;
4.   Approaches based on unit rates.

The most appropriate times for the application of these techniques during the pre-contract phase is shown in Fig. 3.3.

The term 'cost modelling' is now often used to describe the function of cost estimating. The simple fact is that any form of cost prediction can be described as a cost model, whether it be based on functional, performance related, elemental cost analysis or detailed unit rates calculated by contractors when pricing B of Q for tender purposes. However, cost modelling generally implies the use of computer aids in order to allow iterations to be rapidly performed in order to select the most appropriate solution to achieve value for money.

Functional or performance-related estimating

A functional or performance-related estimate typically requires one quantity and one rate and is related to the client's basic requirement. Typical examples include:

- 1000 bed hotel
- 2000 pupil school
- 1500 bed hospital

For example, a hotelier will know that a hotel will cost £50 000 per bed to build and will earn him £50 per bed per night. He can use this information to calculate the relative efficiencies of two proposed hotel options of completely different sizes.

An estimate based on this technique is very simplistic, crude but of course quick. It does not take into account plan shape, number of floors, ground conditions, etc. It is dangerous to use this technique except at the very early stages of inception. Often statistical techniques are employed in an attempt to improve the accuracy and reliability of the estimate.

Size-related estimating

These techniques are invariably based on gross floor area (GFA) approaches when the total floor area of the required building is calculated and then multiplied by an appropriate unit rate per square metre of floor. In former times volumetric approaches were also used but this technique has largely fallen out of favour as large errors can arise.

More detailed approaches can be applied by the use of differential rates for different areas within the building to give a greater degree of accuracy. A major limitation of these techniques is that they take no account of the geometry of the building.

Elemental cost analysis estimating

This technique relies on the selection of one or more suitable cost analyses and adjusting them for time, quantity, quality and location in order to provide an estimate of the building. It is this technique which is used as the means to establish the cost plan which should confirm the budget set at the feasibility stage and to establish a suitable cost distribution within the various elements.

Invariably an outline cost plan is first produced using the cost per square metre of each functional element in order to allocate elemental cost limits. When the design has developed further the elemental unit quantities are calculated in order to establish elemental cost targets for inclusion in the detailed cost plan.

The application of this approach is developed as follows. For an element the most appropriate element unit quantity is measured, for example:

Substructure	Ground floor slab (m^2)
External walls	Area of envelope (m^2)
Stairs	Flights (No.)

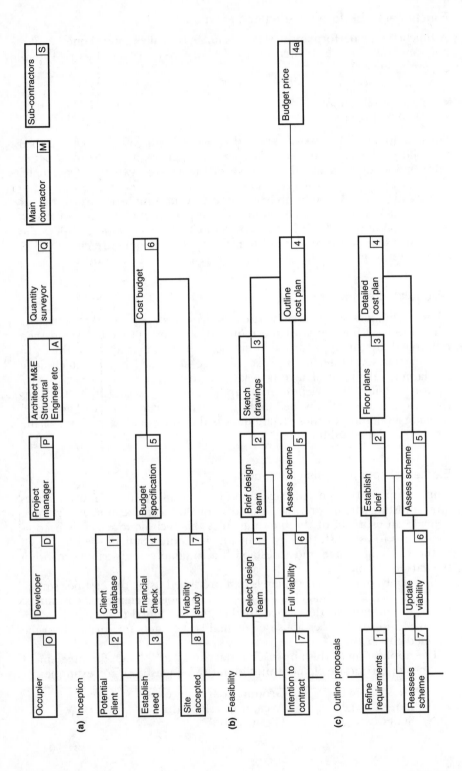

(a) Inception

(b) Feasibility

(c) Outline proposals

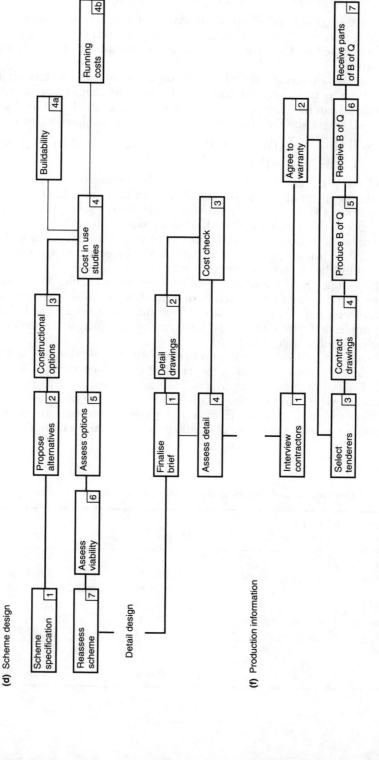

Figure 3.3   Typical construction flow based on the RIBA Plan of Work

Lighting   Gross floor area (m²)
External works   area of site footprint (m²)

In the GFA approach an allowance is made for the stair element without knowing the number, size or quantity. It is very difficult to make any adjustment based on GFA whereas element unit quantities are much more meaningful.

Because the quantities are relatively simple they can be linked; the ground floor plan should be equal to the roof measured flat on plan. Some models can be quite complex, as for example in the Bucknall Austin cost model, where the internal wall finishes area is calculated thus:

$$(EWG \times IFH) + 2\times (IWG \times IFH) - [W + (2\ m^2 \times SED) + (3\ m^2 \times DED)] - [(2\ m^2 \times SID) + (3\ m^2 \times DID)]$$

Where EWG is the external wall girth (m), IFH the internal finishes height (m), IWG the internal wall girth (m), W the windows (m²), SED the single external doors (No.), DED the double external doors (No.), SID the single internal doors (No.) and DID the double internal doors (No.).

Dimensions can be measured for up to fifteen different types of floor plans and each is multiplied by an 'occurrence'. The occurrence is normally one but could be say ten for identical intermediate floors.

As the design evolves more information becomes available. The element unit rate estimate can be modified as described below. Most elements have several different specifications, with varying rates, that need to be isolated. For example, a factory unit may have mainly unfurnished warehouse, some offices and a toilet block.

The element unit rate calculation arrives at the same cost but assumes an identical mix of specification to arrive at an aggregate rate. It is not easy to fully appreciate an aggregate rate as it bears no relation to the specification rates. Any change in the ratios of the varying specification could have a significant cost effect. These parts of elements are referred to as components and are added together to create the elemental sum.

Table 3.1   Establishment of elemental cost targets for inclusion in cost plan

Element 43: Floor finishes Area	Specification	Quantity (m²)		Rate	Cost
Outline cost plan					
All areas	'Typical mix'	10 000		3.65	36 500
		10 000	Total		36 500
Detailed cost plan					
Warehouse	Floor hardener	9 000		1.00	9 000
Office	Carpet	900		25.00	22 500
Toilets	Ceramic tiles	100		50.00	5 000
		10 000	Total		36 500

Cost checking and the use of unit rates

In order to confirm the accuracy of the cost plan, which in itself will have confirmed the budget set at the feasibility stage in the design process, cost checking is deployed. Cost checking is the execution of the cost control component in the design process. It ensures that the information as a basis for the tendering can be prepared such that the lowest tender will confidently equate closely with the budget set at the feasibility stage.

In the cost plan we separated out the different functional areas of the building and applied a rate to the composite construction. The floor finish specification for the office may well read: 'Medium quality floor tiles and hardwood skirtings on raised floor'. In a cost check the carpet tiles, skirtings and raised floor would all be measured separately. The skirtings of course would be measured linearly and would be quite an expensive item. An inclusive area rate would not differentiate between an open plan building versus complex cellular layout, both of the same carpet area but very different skirting lengths.

The same logic applies to other elements and would require the measurement of copings, brick band features, suspended ceiling bulkheads, etc. The cost check also provides a useful quantities check against the B of Q.

Milestone reports

If all the documentation is formatted in the same way it can be compared and reconciled. One way to do this is to use milestone reports. This is a table that summarises and reconciles between each milestone. A milestone is normally a report such as:

- original budget
- cost plans 1 . . . 10
- pre-tender estimates
- contract sum
- financial statement 1 . . . 50
- final account

The main group costs such as total finishes are tabled together with the total cost, area, cost per square metre and a comment on any major changes to the brief. The costs can be plotted on a graph similar to the price risk/design optimisation graph (Fig. 3.4). After several projects the client's cost consultant can analyse his performance to see if he needs to adjust his level of optimism or pessimism!

General comments

The client remembers the first figure reported to him. When the original feasibility study is performed the budget is often cast in stone; it is therefore essential that all cost reporting reconciles back to original budget. All estimates should explain to the client and the design team, what is included in the budget. It should be a discussion document for design optimisation.

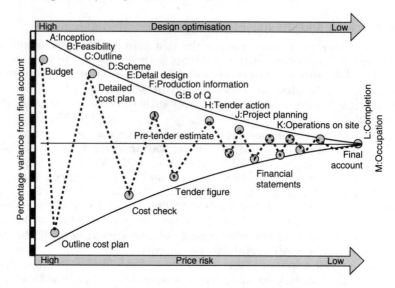

Figure 3.4   Price risk v. design optimisation

As the design develops it is inevitable that some over-specification in individual elements will occur, sometimes increasing the total cost beyond the total budget. The elemental breakdown can highlight the offending elements by showing an excessive percentage of the total.

Normally the individual percentages of each elemental cost for a particular type of building produce a typical 'pattern'. It is important to match the percentages, or pattern, with the norm for that type of building. Mere consideration of the total cost per square metre can be misleading as there can be two high and two low elements which may cancel each other out and yet still require detailed examination.

Care must be taken to ensure that the emphasis on cost control does not cease when the project reaches the B of Q stage, particularly if the bill is reformatted into work sections as opposed to elements. However, computer aids can now play a big role in enabling the rapid reformatting of the B of Q into the most convenient arrangement.

Cost analysis

Types of analysis

Just as there are many techniques to estimating there is an equal number of ways in which to analyse the cost of construction works. Analysis of past projects is particularly important in order to provide data for use in estimating new projects. Either the B of Q or the final account can be used for analysis, there being advantages and disadvantages in using either method.

The B of Q is the most commonly analysed document because the information is readily available and normally formatted in the same way as the cost estimates. Unfortunately most B of Q even though firm lump sums, contain prime cost sums covering nominated subcontractors and suppliers. Further items such as provisional sums, dayworks and the like should be included in the analysis as 'contingencies' unless they can be directly related to a particular element.

All analysis is historical and needs to be adjusted to be appropriate for the current project. Adjustment is done by mathematical formulae to update from the old to the new.

Adjustments to the data

The first adjustment is for 'time', which generally means inflation, but can equally be deflation in a recession. The BCIS published indices can be used to correct historical costs to their projected equivalents. These time indices also include for 'market forces' or profit levels which can often have a more pronounced effect than inflation.

The second adjustment is for 'location'. Again the BCIS publishes indices showing the difference in building costs across the UK. Neither of these indices makes adjustment for 'scale' or 'complexity'. When using analysed data it is important to use a project which is as similar as possible to the current project as the scale and complexity will greatly affect the costs.

In the GFA method of analysis each elemental total is divided by the gross floor area; this item is usually not included as such in the B of Q and may need separate calculation. In contrast the 'functional' unit approach requires that the total cost of the element is divided by the number of functional units.

If the cost analysis can be expanded to include the elemental unit quantities it can be particularly useful in the cost planning process. Care must be taken when analysing not to double count or lose items or include items in the incorrect element; indeed in practice it may desirable to prepare the analysis using the project drawings.

Further adjustments

It is even more useful if the cost analysis can be further broken down to show the detailed breakdown of the various components in terms of quantity and cost of each element. This can be a time-consuming task particularly if carried out manually. Again computer aids can be adopted to carry out this task with both accuracy and speed. Table 3.2 is an example of this approach.

Repetitive projects should also be analysed to show relationships and patterns between dimensions. The obvious dimensions are lettable to gross ratios but also might include:

- window area as a percentage of external envelope
- wall girth to floor area ratio
- car park area to external works area

These ratios can be used for making allowances where quantities cannot be measured or for checking the validity of measured items.

Table 3.2   Cost analysis: internal block work walls

CI/SfB element 22 / BCIS element 2G: Internal walls
Element   22   (CI/SfB): Internal walls   BCIS element 2G

Component: 90   100 mm blockwork walls

SMM item	Quantity Unit	Rate	Extension
Blockwork	100 m²	12.00	1200.0
Extra over fair face	100 m²	1.00	100.00
Ties to steel	25 No.	4.00	100.00
Cutting to soffits	50 m	2.00	100.00
Component quantity total	100 m²	Total	1500.00
Component rate/m²			£15.00/m²

The Building Cost Information Service of the RICS

The nationally available BCIS contains two types of elemental cost analyses namely:

1. Concise cost analysis which gives only a breakdown into broad elements such as substructure, superstructure, services, etc.
2. Detailed cost analysis in a standard format – fully describing each project on five sides, viz:
   (a) information on project, including description, site and market conditions, number and prices of tenders, contract period, form of contract;
   (b) element costs – showing element total cost, cost per square metre GFA, element unit quantity, element unit rate, with two sets of figures for the preliminaries shown separately and apportioned (as Fig. 3.5)
   (c) specification and design notes;
   (d) plan and elevation.

Now the elemental cost data, previously available only in hard copy, is accessible to subscribing members via computer link to the BCIS headquarters in Kingston, Surrey. Thus even in the absence of designer's drawings the client's cost adviser is able to create, on the computer, a pre-contract cost model using the data from several similar previous projects.

Pre-contract cost control on civil engineering projects

Highway works

An approximate estimate of the cost of constructing major highway works is usually required at an early stage in the project cycle in order to

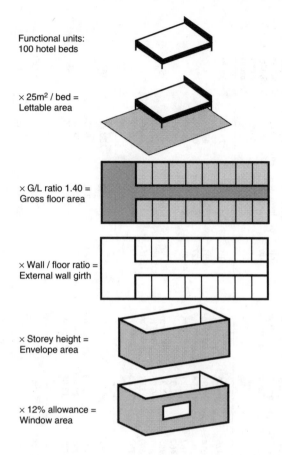

Functional units:
100 hotel beds

× 25m² / bed =
Lettable area

× G/L ratio 1.40 =
Gross floor area

× Wall / floor ratio =
External wall girth

× Storey height =
Envelope area

× 12% allowance =
Window area

Figure 3.5   Fundamental dimension relationships

determine if the scheme is reasonable and will fit into the government funding allowance. At this early stage the proposed project will be analysed in fundamental elements. The road construction will be estimated at £x per linear metre for three-lane, dual or single carriageway – this price will normally include drainage, lighting and signage.

The estimate will be based on an analysis of previous similar tenders using the highway consultant's own data or data from personal contacts, or following up leads in government White Papers or in the *New Civil Engineer*. Adjustments need to be made for inflation and market forces using the Department of Transport's *Road Construction Price Index* which is published quarterly and shows trends in national tender prices.

Additional items to be considered include earthworks, statutory undertakers' equipment to be moved, townscaping and landscaping, telephones, closed circuit TV, etc. Bridges are kept separate and are again estimated based on the consultant's own cost data taking into account the number of spans/type of construction/length/width of carriageway, etc.

Element costs

Base date: June 1990

Gross internal floor area: 15 319

Element	Preliminaries shown separately				Preliminaries apportioned		
	Total cost of element	Cost per m² gross floor area	Element unit quantity	Element unit rate	Total cost of element	Cost per m² gross floor area	Cost per m² at 1985, UK mean location
1 **Substructure**	1 051 478	68.64	7 271 m²	144.61	1 115 416	72.81	54.49
2A Frame	688 014	44.91	15 319 m²	44.91	729 850	47.64	
2B Upper floors	927 704	60.56	8 048 m²	115.27	984 115	64.24	
2C Roof	363 167	23.71	7 271 m²	49.95	385 250	25.15	
2D Stairs	35 467	2.32			37 624	2.46	
2E External walls	448 945	29.31	5 436 m²	82.59	476 244	31.09	
2F Windows and external doors	909 207	59.35	2 936 m²	309.68	964 493	62.96	
2G Internal walls and partitions	361 099	23.57	19 900 m²	18.15	33 056	25.01	
2H Internal doors	435 753	28.45	1 090 No.	399.77	462 250	30.117	
2 **Superstructure**	4 169 356	272.17			4 422 882	288.72	216.07
3A Wall finishes	120 901	7.89	25 006 m²	4.83	128 253	8.37	
3B Floor finishes	405 395	26.46	15 319 m²	26.46	430 046	28.07	
3C Ceiling finishes	199 883	13.05	12 310 m²	16.24	212 037	13.84	
3 **Internal finishes**	726 179	47.40			770 336	50.29	37.63
4 **Fittings**	507 631	33.14			538 499	35.15	26.30
5A Sanitary appliances	236 401	15.43	630 No.	375.24	250 776	16.37	
5B Services equipment		—				—	
5C Disposal installations	261 393	17.06			277 288	18.10	
5D Water installations		—				—	
5E Heat source		—				—	

Figure 3.6   BCIS, hospital project – element costs

Element costs

Gross internal floor area: 15 319

Base date: June 1990

Element		Preliminaries shown separately		Element unit quantity	Element unit rate	Preliminaries apportioned		
		Total cost of element	Cost per m² gross floor area			Total cost of element	Cost per m² gross floor area	Cost per m² at 1985, UK mean location
5F	Space heating and air treatment	3 095 974	202.10			3 284 232	214.39	
5G	Ventilating systems	—	—			—	—	
5H	Electrical installations	1 761 104	114.96			1 868 192	121.95	
5I	Gas installations	—	—			—	—	
5J	Lift and conveyor installations	121 569	7.94			128 961	8.42	
5K	Protective installations	—	—			—	—	
5L	Communications installations	—	—			—	—	
5M	Special installations	—	—			—	—	
5N	Builder's work in connection	276 838	18.07			293 672	19.17	
5O	Builder's profit and attendance	12 990	0.85			13 780	0.90	
5	**Services**	5 766 269	376.41			6 116 901	399.30	298.83
	Building subtotal	12 220 913	797.76			12 964 034	846.27	633.34
6A	Site works	346 048	22.59			367 090	23.96	
6B	Drainage	293 801	19.18			311 666	20.35	
6C	External services	5 179	0.34			5 494	0.36	
6D	Minor building works	437 047	28.53			463 623	30.26	
6	**External works**	1 082 075	70.64			1 147 873	74.93	56.07
7	**Preliminaries**	808 919	52.80			—	—	
	Total (less contingencies)	14 111 907	921.20			14 111 907	921.20	689.41

Figure 3.6 (cont)

If the Department of Transport gives approval to the scheme the highway consultants will then proceed with the detailed design. The initial estimate will then need to be regularly updated – preferably on a monthly basis. The estimate should be continuously refined to include any changes in the scope of the scheme, changes in legislation, e.g. concerning safety barriers or measures to deal with coal seams, new traffic assessment regulations, additional statutory removals, trends in inflation and market forces. Finally when the Department of Transport is fully committed to funding the scheme tenders can sought from selected contractors – normally on a bill of quantities basis prepared in accordance with the Highways Method of Measurement using the ICE Conditions of Contract 5th edition.

The aim of the pre-contract cost management process is for the consultant's estimate to equate to that of the selected contractor; in practice this is not always the case particularly if the estimate has not been updated on a regular basis.

New developments in civil engineering cost management

This chapter has focused primarily on the pre-tender cost management of building projects. This is because the techniques have been specifically developed for the building industry with little regard for the needs of the civil engineering sector. A number of reasons can be addressed as to why this is so amongst them being:

1. The cornerstone of the design cost control in the building industry has been the ability to analyse B of Q into elemental costs as described above and shown in Fig. 3.6. For this to be successful the information in the bill of quantities has to be prepared in a logical and consistent manner. This has long been achieved in the building sector through the use of the various standard methods of measurement prepared by the RICA and the BEC (formerly the NFBTE) on behalf of the building industry. Until 1976 the standard method of measurement used by the civil engineering industry was somewhat vague in its content and therefore subject to various interpretations thus leading to inconsistencies. In 1976 a completely revised standard method of measurement (now in its third edition) was introduced which eliminated many of the problems associated with the earlier edition, thus affording more opportunity for the application of approaches developed in the building sector.
2. Civil engineering projects are generally of a much more 'one-off' nature due to the uniqueness of many such projects in terms of their design and the fact that they often involve major groundworks and similar uncertain conditions.
3. The fact that quantity surveyors have not often been involved in the pre-contract stage of civil engineering projects and therefore their techniques and expertise on pre-tender cost control have not been deployed.

4. Most civil engineering undergraduate programmes are not able to consider the use of cost management techniques in any depth in their syllabuses because of the importance of other aspects of civil engineering education.

However, the Water Research Centre have built a number of cost models based on statistical techniques using regression analysis. These techniques are documented in their Technical Report TR61 produced in 1977 and have recently been updated in their 1992 report. Unfortunately this latest report is not available in the public domain as it remains confidential to the various water authorities who have invested in its production. The aim of the work carried out by the Water Research Centre is to assist in capital investment planning and estimating the costs of specific schemes.

Pre-contract estimates can now be produced for civil engineering works on an elemental basis following the development of the system by North West Water, Liverpool Polytechnic (now Liverpool John Moores University) and the BCIS. This approach has now been included in ICEMATE – a civil engineering estimating computer software package developed by Thomas Telford Services and CCSP software. The computerised system allows for input of elemental or unit price information from previous projects and permits the automatic transfer of data from the civil engineering bill preparation package ICEPAC.

This approach is currently being further developed by the BCIS and Liverpool John Moores University who are researching into the development of a standard form of cost analysis for civil engineering work. If this proves successful it should provide the civil engineering industry with a nationally available approach to analyse, in a consistent and logical manner, the costs of civil engineering projects. This should help to provide more reliable and objective pre-contract cost management of such projects.

Further developments in design cost control

Value management is an organised and systematic approach to obtain the best functional balance between cost, reliability and performance. Its objective is to identify key areas of unnecessary cost and to seek new creative ways of obtaining the same performance more economically.

Although most effectively applied in the early stages of a project, a value management study can still identify potential savings later in the project cycle. However, the opportunity for effective cost reduction is much greater earlier in the project cycle – individual changes generally cost the client more later in the project cycle.

The value management approach encourages better communication at the design stage and a greater integration of the design. The system revolves around a planned series of highly structured 'think-tank' sessions chaired by a professional facilitator. Typically these sessions could take place at the following stages: feasibility, concept design, design development, appointment of contractor and post-occupancy.

It is claimed that value management, which has been standard practice in the USA and Japan for decades, can save up to 25 per cent of the cost on some major projects.

Tutorial questions

1. Describe and discuss the range of cost models and show how they are useful at different stages in the design process. (RICS Direct Membership (Quantity Surveying), Project Cost Management, 1989)
2. Discuss the steps that can be taken to ensure that cost planning and control keep the final cost of a building project within the cost target. (CIOB Member Examination, Part II, Contract Administration, 1990)
3. The primary function of producing estimates of the cost of construction works is to be able to advise clients of anticipated development costs. Discuss the various methods of providing such pre-contract estimates in relation to a proposed marina development. (RICS Final Examination (Quantity Surveying), 1991)

Further reading

Ashworth A 1983 *Building Economics and Cost Control: Worked Solutions* Butterworths
Ashworth A 1988 *Cost Studies of Buildings* Longman Scientific & Technical
Ferry D J, Brandon P 1991 *Cost Planning of Buildings* 6th edn BSP Professional Books
ICEMATE, ICEPAC, Thomas Telford software (tel: 071 987 6999)
Kelly J, Male S 1993 *Value Management in Design and Construction* Spon
Raftery J 1991 *Principles of Building Economics: An introduction*, BSP Professional
Seeley I 1983 *Building Economics* 3rd edn Macmillan Press
Skitmore M 1991 *Early Stage Construction Forecasting: A Review of Performance* Occasional paper, RICS Books

4

Financing of major projects

Introduction

The arrangement of funds to finance the development and construction of a major project will vary from project to project depending on the type of project and the location. Projects could be funded from many diverse sources, e.g. by governments, possibly aided by the World Bank or some other development agency, by banks and other financial institutions or by the major corporations themselves.

However, the great majority of finance for major public sector infrastructure construction projects, in the UK and abroad, still comes from government funds raised through borrowing or the general government revenue.

The financing of major private sector construction works is quite different from the loan given to a financially sound company as part of its continuing investment programme. The analysis of the loan proposal for the project will involve a credit decision based on the review of the project, particularly the risk involved, rather than upon historical earnings.

Experience indicates that serious problems are frequently encountered on major construction works. Typical risks to which the financiers might be exposed include:

1. Completion on time and within cost.
2. Operating risk: when complete the project may not perform as anticipated resulting in a higher operating cost or a lower productivity.
3. Market risk: by the time the project comes on stream the demand for the product may be reduced, possibly due to the completion of competing products or reduced demand.
4. Reserve risks: the cost of recovering the natural resources e.g. oil, gas, iron ore or coal, may prove too great or the amount of the reserves may have been overstated.
5. Political risk: changes in taxation, import duties and environmental legislation may have serious impact on the project.
6. Technology risk: technology used becomes obsolete or cost ineffective even before completion.

Types of finance

There are a wide variety of funds generally available for the construction of major projects from a range of lenders and investors. In practice most projects are financed using a variety of methods utilising several different sources.

Equity capital

Raising money through the issuing of shares traded on the Stock Exchange is a popular way of financing major projects. In return for his/her investment the investor will become an owner of the company and be able to vote on matters of policy and will receive a share of the company profits in the annual dividends.

Loan capital

If a company does not wish to, or cannot, obtain long-term financing from its shareholders it must borrow from outsiders. Such borrowings are called loan capital or medium and long-term debt.

Loans to companies are sometimes in the form of debentures issued at face, discount or premium value depending on the company's circumstances at the time of issue. Holders of debentures have the highest security within the company and are guaranteed a fixed interest whether or not the company is making a profit.

Bank loans or an overdraft facility can provide a means of raising finance for capital projects. However, increasingly larger construction projects require a greater proportion of outside finance. The leading commercial banks are now taking an active role in the financing of major projects – undertaking financial feasibility studies and making recommendations – the Channel Tunnel project being a prime example.

Commercial paper

Commercial paper, which can be issued in any currency but mainly in sterling, US$ or Eurodollars, may also provide an important source of finance for major projects. Commercial paper is in fact promissory notes issued through agent banks who in turn place with investors who may be banks, insurance companies, provident funds, etc.

Non-recourse financing

In this type of financing the lender must look to the project itself to provide the security for repayment. This arrangement has been used for many years to finance major North Sea oil developments and has been utilised to finance the building of power generation plants by the privatised electricity companies.

The proposal for the loan, known as the 'information memorandum', is prepared by a merchant bank and then offered to other banks and finance houses. The loan is often structured to include: senior and subordinated debts, equity, bond and capital leases. Advantages of this arrangement include: the company or joint venture can increase its borrowing capacity beyond its normal resources and the entire project can appear off-balance sheet with possible tax benefits.

Disadvantages of non-recourse financing include the fact that interest rates may be higher than conventional arrangements – reflecting the greater risk to the lender when project revenues are locked into a scheme during the period of financing. If the banks seek to minimise their exposure they may insist on the developer guaranteeing completion dates or final project costs – known as limited recourse financing.

Public grants

Many sources of grants are available to developers and local authorities within the UK. These grants, loans and incentives can be particularly relevant in the case of socially desirable but commercially unsound projects. However the ground rules for receipt of these grants are constantly changing and project managers should check the latest position with the appropriate authority.

Sources of grants available from the European Union (EU) include:

- *European Investment Bank (EIB)*. The EIB was established by the Treaty of Rome with European Community (EC) member countries subscribing capital in rough proportion to the size of their economies. Loans of up to 50 per cent are provided for both private and public sectors for investment in industry, infrastructure and energy.
- *European Regional Development Funds*. Grants, up to 50 per cent in value, are available for industry and infrastructure projects to public authorities and the private sector in assisted areas as defined by the Department of Trade and Industry. In practice the emphasis is on the creation of jobs in those areas affected by industrial decline.

The Commission of the European Communities also provides loans of up to 20 per cent for nuclear projects, whilst the European Coal and Steel Community provides loans for projects in coal or steel or for assistance in closure/run-down projects.

Amongst the many sources of grants available within the UK are:

- the city grant which benefits projects in run-down urban areas which could not proceed without public support;
- the Urban Programme under which the government makes available a 75 per cent grant to local authorities towards the development of inner cities;
- derelict land grants which make up to 100 per cent available to local authorities (and up to 80 per cent to private developers in assisted areas) for the regeneration of derelict land.

Sources of finance

Within the UK loans are normally obtained from commercial banks, finance houses, building societies and other financial institutions, e.g.

pension funds. Construction companies themselves may finance projects utilising their working capital. Property and development companies may also be prepared to finance, or part finance, a major development in return for equity in the completed scheme.

Additional funds can also be raised for large capital projects from the Eurocurrency market, either directly from a limited number of international banks or through a syndicated loan arranged with a group of banks. The terms of these loans can be extremely flexible and can be arranged in any currency (although Eurodollar loans predominate). However, loans from the Eurocurrency market are usually more expensive in terms of fees and charges.

The international bond markets in Europe, the USA and the Far East also provide an additional source of funding, often at favourable interest rates, for major capital projects. These markets offer flexibility with fixed or variable rate funding with a range of currencies and maturities.

The New York commercial paper market also provides short-term financing to major multinational corporations who have been accorded a top credit rating by US rating agencies. Funds are raised through issuing and selling commercial paper, on average at lower funds than those applying in the Euromarket.

Build–operate and transfer (BOT) contracts

Until quite recently the majority of large construction projects, with the exception of those in the oil, gas and process plant sector, had been financed by governments. However, many of the UK government-sponsored projects, undertaken for the public good, suffered time and cost overruns. Furthermore in the 1980s international banks suffered severe financial losses in various Latin American and other developing countries resulting in the withdrawal of medium and long-term bank loans to these countries.

The time was right for the reintroduction of the BOT (build–operate and transfer) concept for the construction of major infrastructure projects in both developed and developing countries. The concept was not new and had first been used by the Brothers Perrier on a water supply project to Paris in 1850. In 1984 it was taken up by the Turkish government on the privatisation of public projects.

Many major worldwide infrastructure projects have since been funded under the BOT arrangement including: Sydney Harbour Tunnel (Australia) US$550m. (Fig. 4.1), Shajiao Power Plant (China) US$517m., Dartford Bridge (England) US$310m., Eastern Harbour Crossing (Hong Kong) US$600m., Second Stage Expressway (Thailand) US$880m., North–South Expressway (Malaysia) US$1.8bn. Under this arrangement a private organisation offers to build and operate a project and then return the ownership to the government after a fixed concession period.

The main parties in a BOT arrangement include: the host government, the project sponsors, the contractors, the banks and the shareholders. The investors in the project look to the revenue generated from the completed scheme to repay debts and provide a profit on their investment.

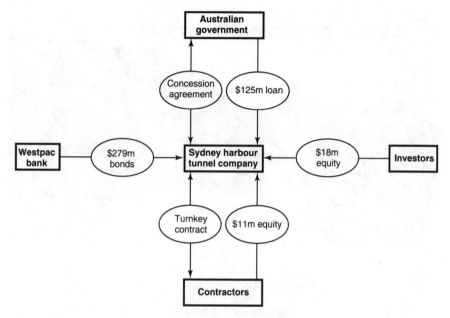

Figure 4.1 Project structure of Sydney Harbour Tunnel. *Source*: Tiong[1]. Reproduced by permission of the American Society of Civil Engineers.

The host government may invite bids from interested sponsors or may receive speculative bids from individual sponsors. The sponsors, who are usually a major construction contractor or a consortium of contractors, have a wide role and responsibility under this arrangement including:

- carrying out the feasibility study
- carrying out the engineering design
- negotiating favourable concession agreement with the government
- raising the equity to fund the construction of the project
- building the project
- operating the project

BOT projects usually involve a high degree of risk and uncertainty. The concession agreement signed by the host government and the project sponsor defines the scope of the project in commercial terms and allocates the political, legal, commercial and environmental risks. Concession agreements may be statutory or contractual or a combination of both.

Tiong[2] identified that the host government, whilst not providing direct funding to major BOT projects, did provide certain guarantees and incentives which proved critical in attracting finance for the projects. Such guarantees included: fixed concession periods (between 10 and 55 years), support loans, guarantee of minimum operating income, concession to operate existing facility, commercial freedom, foreign exchange guarantee, interest rate guarantee and 'no second facility' guarantee.

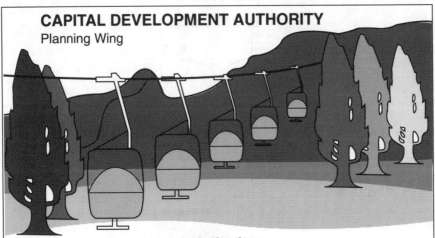

CAPITAL DEVELOPMENT AUTHORITY
Planning Wing

Invitation for
EXPRESSION OF INTEREST
for INSTALLATION and MAINTENANCE
of CHAIRLIFT PROJECTS
at MURREE/ISLAMABAD

In order to promote and encourage tourism and recreational facilities in Islamabad and adjoining Murree hills, CDA intends to build chairlift facilities at following locations.

S.NO.	LOCATION	Reduced Level at Base Station	Reduced Level at End Station	Approx. Distance in feet
MURREE				
1.	Chitta Mor - Mall	6175 Ft.	7356 Ft.	2640 Rft.
2.	Jhika Gali - Mall	6400 Ft.	7356 Ft.	2300 Rft.
ISLAMABAD				
1.	Faisal Mosque - Pir Sohawa	4200 Ft.	1940 Ft.	25500 Rft.
Alternatively	Noorpur Shahan Village - Pir Sohawa	4200 Ft.	2000 Ft.	12600 Rft.
2.	Shakarparian Hill - Rawal Lake Hill	1925 Ft.	1818 Ft.	16000 Rft.

2.   Apart from Chairlift facilities, Children Playland, Motel/Restaurants and other amusement facilities may also form part of the development. All these locations have been selected with respect to their commercial sensitivity and viability. A suitably worked out business plan could easily fetch a high return on investment. Government is very keen to develop this whole area into a popular touristic spot not only for local but also for touristic traffic from neighbouring countries.

3.   The CDA undertake to provide following facilities:–

3.1 Lease of Land to party for agreed period of time on amenity rate.

3.2 Construct and develop associated infra-structural works including approach roads, electricity, parking etc.

3.3 Facilitate for safe return of the investment.

4.   Interested parties are invited to submit their expression of interest in putting a proposal, in collaboration with equipment manufacturers to provide, install and subsequently operate the proposed Chairlift projects on turn key basis with or without financing. Comprehensive expression of interest accompanied by complete resume of the firm(s) should reach the undersigned by March 15, 1993. Parties having experience in similar work on mix financing or BOOT or BOT or BOO basis should specify such projects in particular with details.

Deputy Director General
Planning CDA.

Figure 4.2   BOT contract – invitation of expression of interest. Source: *Daily Telegraph*, 2.2.93

The benefits of the BOT system include minimising of state borrowing, the potentially greater efficiency of private enterprise and the stricter financial discipline which is imposed on the lending institutions who must rely upon the success of the project for a return on their investment.

The BOT arrangement has enabled funds to be raised to finance industrial as well as infrastructure projects. It may also be appropriate for other types of projects, e.g. prisons, schools, hospitals, sheltered housing, halls of residence, waste disposal plants and car parks.

The Hong Kong Mass Transit Railway Corporation (HKMTRC) were first involved in a BOT contract on the US$600m. immersed tube project under the eastern harbour from Kowloon to the Hong Kong island. Under the arrangement the New Hong Kong Tunnel Company Ltd (a consortium headed by the major Japanese contractor Kumagai Gumi) entered into a BOT contract with the Hong Kong government to build a combined rail and road crossing.

The consortium obtained a franchise allowing them to operate the road tunnel until 2016 and the rail tunnel until 2008. The tolls collected and the money received from the rail lease (from the HKMTRC) would provide the payback for the whole project which opened in September 1990 – four months ahead of schedule. Approximately 20 per cent of the finance for the project was raised by the consortium's shareholders with the balance arranged by Shearson Lehman Brothers Inc. utilising funds from 50 banks around the world.

For further details of this and other BOT projects involving Japanese contractors see Levy's *Japan's Big Six*.[3]

Case study No. 1 – UK BOT contract – the Channel Tunnel

The Channel rail tunnel between the UK and France is the largest BOT project attempted anywhere in the world. The tunnel will be operated by Eurotunnel, which is a private sector group whose shares are listed in London and Paris, under 'the Concession' for a period ending in 2042.

Eurotunnel assumed full construction risks for the Channel Tunnel and arranged a £1.1bn. funding margin for cost overruns. Eurotunnel awarded the construction contracts to its contractor – who were also the founder shareholders – Transmanche Link (TML). The contractor had an obligation to design, construct, test and commission the system.

The tunnel work itself was on a target cost contract with Eurotunnel agreeing to pay the contractor the actual costs plus 12.36 per cent of the target value. If the costs exceeded the target TML were to bear 30 per cent of the costs above this level, if the costs were less than the target there would be a 50/50 split of the saving. In addition, contracts were subject to price adjustment for unforeseen ground conditions, variations and inflation.

Sources of finance

In 1986 the Channel Tunnel Group and France Manche (later to become Eurotunnel) won the concession for the construction of the Channel

Tunnel. The group comprised ten construction companies (five British and five French) and five banks (two British and three French).

By September 1986 the group had 'kick-started' the project and had raised £46m. in equity funding. In October 1986 a further £206m. was raised by a private institutional placing. At this stage Eurotunnel had a construction contract essentially written by the contractors themselves and approved by the banks!

Preliminary work on the project began and in November 1987 Eurotunnel raised a further £770m. with a public share issue. A rights issue raised a further £566m. Under this arrangement existing shareholders were given the option to buy further shares below the current market price, the number of shares offered being proportional to the number of shares held. Further equity of £101m. arose in 1992 from the exercise of the Equity 3 warrants.

In October 1990 Eurotunnel entered into the Revised Credit Agreement with an international group of lending banks increasing the total of the project loan from £5000m. to £6800m. These credit facilities are required to be repaid from Eurotunnel's operating cash flow. The EIB and Crédit National agreements provided credit lines of £1000m. and 4000m. Fr. respectively which are supported by letters of credit (shown as interest guarantees of £78m. on the sources of funding statement). The stringent conditions attached to the Revised Credit Agreement, which extended the final maturity of the facility to the year 2012, enabled the banks to monitor the progress of the project and the expected cash flow and to exercise control in the event of any significant adverse developments.

In May 1990 Eurotunnel obtained a further £300m. on a 25-year fixed rate of interest from the EIB. The Luxembourg-based EIB was the largest single lender to Eurotunnel with a total commitment of £1300m.

The financing requirement for the project includes all construction and corporate costs estimated to be incurred to the end of the construction period, together with associated financing costs and provision for inflation. Thereafter the financing requirement has to cover a short period during which the net operating revenues are exceeded by capital expenditure and debt service requirements.

The total equity and credit facilities shown in Table 4.1 is £1104m. more than the total estimated cash requirement to address the possibility of additional costs. These additional costs could be: construction claims made by TML, net cash outflow as a result of loss of revenue due to a late opening or slower than expected traffic growth, financing through increased interest rates or excessive inflation.

In January 1994 Eurotunnel was given an extra 10 years' lease to the 55-year facility by the British and French governments; this was in return for Eurotunnel withdrawing contractual claims against the two governments. The extension of the lease to the year 2052 assisted Eurotunnel to raise an additional £1bn. needed to complete the link and pay for the additional safety, security and environmental measures; half the money was raised through a rights issue and half through the banks.

Table 4.1   Sources of finance for the Channel Tunnel (£m.)

Equity			
Equity 1	46		
Equity 2	206		
Equity 3	770		
	———	1203	
Rights issue		566	
Exercise of Equity 3 warrants		101	
		———	1690
Credit facilities			
Main facilities			6800
European Investment Bank			300
Total financing			8790
Interest guarantees			(78)
Total cash resources			8712

Source: Eurotunnel Rights Issue November 1990.[4] Reproduced by permission of Eurotunnel

Overseas projects

Promoters for overseas projects may be able to raise finance by utilising two additional sources of finance, namely: the international development banks and the international export credit agencies.

International development banks

The UK contributes to a wide range of international organisations which then finance their own programmes in developing countries. The Overseas Development Administration (ODA) supports programmes funded by the European Union (EU), the World Bank, the United Nations agencies and the Regional Development Banks (Inter-American, Asian and African).

The World Bank lends finance for development projects which contribute to the development of the country's economy. Since opening its doors in 1946, the bank through the International Bank for Reconstruction and Development (IBRD) and its soft loan affiliate, the International Development Association (IDA), has made over 3000 development loans and credits for a total of US$100bn.

Loans can be made by the World Bank to member governments, public agencies or corporations, or private bodies with the government's guarantee. Increasingly lending has been directed towards poor and less developed countries in Asia, Africa and Latin America. Generally the bank lends for specific projects such as schools, crop production programmes, hydroelectric power dams, roads and fertiliser plants.

The opportunities for UK companies and consultancies to do business through multilateral funded projects are substantial and are not limited by the size of the British contribution to the organisations concerned. Indeed it is estimated that the UK receives orders for British goods and services to the value of over one and a half times its contributions to multilateral aid agencies.

Export credit agencies

Most industrialised nations have government-supported export credit agencies which provide short- and medium-term export credit insurance and guarantee schemes. Many of these are arranged in accordance with an agreement, intended to prevent harmful competition, made under the auspices of the Organisation for Economic Co-operation and Development (OECD). Twenty-four nations accepted the agreement or 'consensus terms' including members of the European Union and Japan.

British overseas aid

The ODA is the UK government department responsible for aid to developing countries. The aim of the aid is to promote economic and social development and to alleviate poverty. Aid also opens up opportunities for British contractors, suppliers and consultants in projects financed by the ODA directly. However, aid cannot be used to subsidise private sector ventures but it can be used to assist or complement private investment by financing associated infrastructure or through training or consultancy.

Bilateral aid accounts for about 60 per cent of total British aid and is provided on a country-to-country basis mainly to the poorest ex-Commonwealth countries. As a general policy the UK government require that goods and services provided under bilateral aid should be of British origin.

Bilateral aid may help fund new capital investment schemes like power stations, roads, railways or water supplies. It may also take the form of programme aid, e.g. raw materials or spares to keep existing investments going. The aid may also be in manpower, for example UK consultants involved in preparing feasibility studies/cost–benefit analyses/environmental impact studies for the project, or professionals from the developing countries visiting the UK for specialised training.

A small proportion of the UK bilateral aid funds is set aside for 'aid and trade provision' (ATP). These funds are allocated in grant form or long-term low-interest rate loans in support of sound projects of particular commercial and industrial importance to the UK. The request for the ATP, which are normally associated with export credits guaranteed by the Export Credit Guarantee Department (ECGD), comes initially from the British company (via the Department of Trade and Industry) rather than from the recipient government (Table 4.2).

The UK's bilateral aid programme is also supported by the Common-wealth Development Corporation (CDC) which supports a wide range of

Table 4.2   Examples of ATP supporting UK companies

Country	Project	Company	UK export value (£m.)
China	Yueyang Power Station	GEC	Over 170
India	Balco Power Station	GEC	130
Indonesia	Cigading Port	Tarmac Consortium	17
Turkey	Ankara natural gas	AMEC, British Gas	70
Colombia	System X	Plessey (now GPT)	10
Kenya	Rural tele-communications	Communications supplies	25

Source: *British Overseas Administration: Opportunities for Business*, Overseas Development Administration.[5] Reproduced by permission of the Overseas Development Administration

enterprises both public and private in the developing world. The CDC is an investment institution funded by loans from the aid programme and by self-generated funds.

The export credit or insurance agency will normally insure the exporter against political risks, together with some of the commercial risks relating to the buyer, and will provide a guarantee of repayment to the lending bank.

The main advantages of export credit (buyer credit or supplier credit) lie in its low fixed interest rates and long repayment terms. Typically a promoter's export credits could carry fixed interest rates between 7.5 and 9.25 per cent with repayment periods ranging from 8.5 years to 12 years after commissioning.

In buyer credit, finance is usually arranged and provided by a merchant bank working in conjunction with the appropriate export credit authority. The merchant bank, either alone or with partners, provides finance to the buyer (the promoter) for payments on work completed and the promoter undertakes to repay the loan to the lender (merchant bank) on fixed repayment terms as agreed.

In contrast under supplier credit the supplier (contractor), after negotiating with the export credit authority for the best terms possible, acts as the lender. When the contractor submits a tender it will include the detailed terms of export credit finance to make the offer more attractive. Payments for work completed are deferred and become part of the export credit loan.

The £2bn. Cairo wastewater scheme is one of the most recent projects to which the ODA has contributed. The major rehabilitation of the Cairo sewerage system involved constructing 50 km of tunnels, pumping stations, culverts and a major treatment plant.

The ODA provided a £66m. grant which together with a £185m. loan from the Midland Bank backed by the ECGD helped the Egyptian government fund the project. The first phase was completed in 1992 under ten separate contracts valued at £500m.; most of the contracts were carried out by joint ventures involving leading Egyptian and British companies.

Case study No. 2 – the Hong Kong Mass Transit Railway

Construction on the first stage of this £2000m. underground transport system commenced in 1975 with phase three being completed in 1984. The raising of the finance for this massive infrastructure project was an extremely challenging and at times difficult exercise for the corporation and its financiers.

Initially the corporation – a newly incorporated body unfamiliar with international finance – adopted a conservative financial strategy. The costs of civils and electrical and mechanical (E&M) construction together with interest payments and finance for phase one of the project were raised on the open market, through export credits and local and overseas loans – mostly carrying Hong Kong government guarantees. A further sum was provided by the Hong Kong government for land costs, administration and consultants' fees in return for equity in the corporation.

Export credits were one of the most important sources of finance for the HKMTR providing 34 per cent of the total construction cost of the first three phases of the project up to 1984. Most of the contracts awarded to the UK or French contractors had provisions for buyer credit whilst the contracts awarded to the Japanese contractors carried a supplier export credit facility. Furthermore, in contrast to the first stage in which the UK was the main provider of export credits, the Japanese eventually emerged as the main provider of this sort of funding over the first three stages.

Market finance provided the other important source of funds accounting for about 44 per cent of the total financing up to 1984. The basic strategy was aimed at a mix of various debt instruments, direct and syndicated loans and short and medium term facilities.

Property developments along the route of the metro, mainly above underground stations, were the third important source of revenue providing 10 per cent towards the cost of construction up to 1984.

However investing in property is a risky business particularly in a depressed market. The poor market conditions in Hong Kong in the early 1980s necessitated a revision to the Corporation's strategy with additional loans being required to be raised from the market with an extension to the final payback of three years.

The fourth source of finance up to 1984 for the Corporation was equity received from the HK Government to the value of HK$5500m. of which part was equity issues in lieu of payment for land premium and rates.

Financing stage one

The finance sources for phase one (Modified Initial System) were fully booked up front to ensure the availability of funds. Lines of credit were negotiated with various export credit authorities like ECGD, or similar overseas financial bodies, to enable the contractors from these countries to compete for contracts. Eventually, however, only 30 per cent of the total requirement for phase one was covered by export credit with the UK providing half the total.

In those early days the corporation was forced to accept various currencies, other than HK$, which were obviously high risk in times of a fluctuating HK$. Commercial loans were also arranged up front, both in US$ and local currency. Very high financial charges were incurred in the form of high commitment fees as well as interest with margins running as high as 2.5 per cent.

Stage two

Construction of phase two of the project (Twuen Wan Extension) commenced in 1978. The most notable shift of policy was in reducing foreign currency borrowings with over 97 per cent of export credit lines being negotiated in HK$. Furthermore all syndicated US$ loans were either prepaid or converted into HK$. The corporation also became interest sensitive and began using short-term money facilities.

Stage three

On the third stage of the project (Island Line), all sources of finance were denominated in local currency, whether they be export credits, short or medium term; this proved to be a most prudent policy with the weakening of the HK$ in the second half of 1983. As the corporation grew in popularity it succeeded in obtaining finer margins, low commitment fees, longer periods of grace and improved documentary terms.

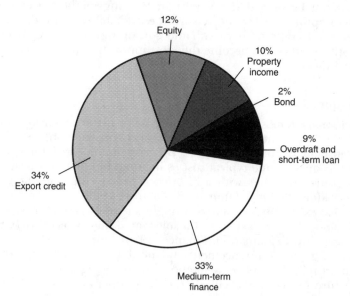

Figure 4.3 Sources of finance for the first three stages of the Hong Kong Mass Transit Railway (up to December 1983 prior to syndicated loan). *Source:* Lau[6]. Reproduced by permission of the Hong Kong Mass Transit Railway, and W. S. Lau

However, all was not plain sailing and in 1982 the collapse of the Hong Kong property market and political uncertainties led to revision of strategies. The previous funding horizon of 9 months was extended to 18 months and a major syndicated loan of HK$2000m. consisting of two tranches with option to double the loan was arranged at the end of 1983.

Despite the heavy debt burden all the borrowings for the first three stages of this mammoth project were anticipated to be fully repaid to the HK government by 1996.

The future

The preferred financial model for the funding of the new 34 km Hong Kong Airport Railway scheme due to commence construction in 1993 would show 25–50 per cent raised from the capital markets, 0–10 per cent from export credit and 40–60 per cent from medium term in various categories. In terms of the markets, 20–30 per cent from Japan, 20–30 per cent from the USA, 10–20 per cent from Europe and 30–50 per cent from Hong Kong.

The future financing programme of the corporation will include export credit loans, private placement US dollar bonds and public yankee bond issues, US medium-term note programme, Samurai yen bond and private placement 'Shibosai' bonds, Eurocurrency issues and Hong Kong dollar bond issues. These will be supplemented by the corporation's existing European Monetary programme, and the commercial paper programmes in the US, Euro and Hong Kong markets.

Thus it can be seen that a variety of funding methods were and will be used to optimise costs and ensure the availability of funds. Strategies were reviewed continuously, with the corporation changing from using conservative methods to become the most innovative and pioneering body in Hong Kong.

Tutorial questions

1. A national processing company proposes a major extension consisting of heavy civil, mechanical and electrical construction, together with the installation of major plant. Describe methods of financing such projects and outline typical cost estimating, bidding and post-contract procedures together with appropriate procurement methods. (RICS Direct Membership Examination, Finance, 1985)
2. To what extent and in what way might a major contractor be prepared to finance the contract cost of a substantial new project? (RICS Direct Membership Examination, Finance, 1986)
3. Following the announcement by the UK Chancellor in the 1992 autumn mini-budget, new methods of financing private sector infra-structure projects, particularly the build–operate and transfer (BOT) method, have become more significant.
   (a) Describe how such a system operates in practice and identify the main parties.
   (b) What are the roles and responsibilities of the sponsoring contractor?

(c) What are the main risks and possible solutions carried by the project sponsors during the construction and operational phases?
(d) What types of projects are suitable for this arrangement?

(University of Wolverhampton, Construction degree programme, Examination, Part 3, Finance, 1993)
4. Compare and contrast the role of the financial system in Japan with that encountered in the UK

References

1. Tiong R L K 1990 Comparative study of BOT projects. *Journal of Management in Engineering* **6** (1), January: 107–22
2. Tiong R L K 1990 BOT projects: risks and securities. *Construction Management and Economics* **8**: 315–28
3. Levy S M 1993 *Japan's Big Six: Inside Japan's Construction Industry* McGraw-Hill
4. Eurotunnel rights issue, November 1990
5. *British Overseas Aid: Opportunities for Business* Overseas Development Administration, Information Department, 94 Victoria Street, London SW1E 5JL
6. Lau W S Financing the Hong Kong Mass Transit Railway (unpublished paper)

Further reading

Barnes M (ed) 1990 Financial control. In Jackson M, Richardson M (eds) *Financing Construction Projects* Thomas Telford ch 4
Baum W C 1989 *The Project Cycle* The International Bank for Reconstruction and Development/The World Bank, December
Breakthrough – the story behind Eurotunnel's Funding. A sponsored supplement reprinted from *Euromoney*, March 1991
Bull M P, Savage K H Financing the project. In Lock D (ed) *Project Management* Gower, pp 169–87
Butler E, Moore P 1989 Sources of funding public sector support. *Architects' Journal* 16 August: 53–66
Castle G R 1975 Project financing – guidelines for the commercial banker. *The Journal of Commercial Bank Lending* April: 14–30
Clark M, Tinsley M 1991 Project finance. In *Guidelines for the Management of Major Construction Projects* NEDO, HMSO pp 23–9
Maddox E, Potts K F 1993 The financing of major power projects within the UK. *Civil Engineering Surveyor* March: 9,10
Merna A, Smith N J The allocation of procurement risks in BOT contracts (unpublished)
Moss C R 1992 The Capital Markets Conference 1992, IIR Conference, 14 September
Stannard C J 1984 Let's hand project management over to the banks. *Proc. Instn Civ. Engrs* Part 1, **76** (August): 774–8

Section C

Contract strategies

5

Organisational method

Introduction

A promoter is required to make fundamental decisions at two important stages in the project cycle: at the *appraisal* stage and at the *contract strategy* stage.

At the appraisal stage, the promoter guided by his professional advisers will usually assess the commercial prospects of the project, taking into account the major risks that may affect the project, before deciding whether to proceed. Following the decision to proceed, the promoter must then determine which contract strategy is most likely to effect the change from conception to realisation of the project. Contract strategy thus covers the overall pattern of decisions made by the promoter in defining the organisation and procedures required for the execution of a project throughout the design, construction and commissioning periods.

The strategy chosen will be governed by consideration of priorities, including speed, degree of complexity, quality, flexibility, competition, price certainty, incentive, responsibilities and risk sharing. The importance of choosing the appropriate strategy for the particular project should not be underestimated: 'In essence procurement advice can have an equally dramatic impact on a project's performance when measured in terms of cost and time as can design and engineering advice on a project's performance when measured in terms of function and quality.' (Alan Yates, Dearle and Henderson)[1] However, it is also equally true that at the end of the day the professional approach of client and the attitude of the designers, the project management team and the contractor can have an equally important influence on the project success.

The following four subsections need to be considered when developing a strategy for a project:

1. Organisational method: e.g. traditional, design and build, management, turnkey, BOT, partnering.
2. Type of contract: e.g. lump sum, admeasurement, cost reimbursable, target.
3. Bidding procedures: e.g. open, selective, two-stage, negotiated.

4. Conditions of contract: e.g. JCT 80, JCT 87, ICE 6th, FIDIC, GC/Works 1 (Edition 3)

Recent trends

The 1993 biennial survey undertaken by the RICS showed the following procurement methods in use in the building sector in 1991 in the UK (Table 5.1). The survey did not include overseas work, civil engineering, heavy engineering, term contracts, maintenance and repair contracts and subcontracts.[2]

Table 5.1   RICS survey showing trends in procurement methods in the building sector. *Source* Reference 2. Reproduced by permission of the Royal Institution of Chartered Surveyors

Procurement method	*Percentage of contracts by value*				
	1985	*1987*	*1989*	*1991*	*1993*
Lump sum – firm B of Q	59.26	52.07	52.29	48.26	41.63
Lump sum – specification and drawings	10.20	17.76	10.26	8.35	9.98
Lump sum - design and build	8.05	12.16	10.87	14.78	35.70
Remeasurement – approximate B of Q	5.44	3.43	3.58	1.26	2.43
Prime cost plus fixed fee	2.65	5.17	1.12	0.12	0.15
Management contract	14.40	9.41	14.99	7.87	6.17
Construction management	—	—	6.89	19.36	3.94
Total	100.00	100.00	100.00	100.00	100.00

Table 5.1 clearly shows the trends within the building sector in recent years: there has been a dramatic decline in lump sum contracts – with or without quantities. Prime cost contracts and management contracts have also declined in use whilst design and build has gained popularity.

The Ministry of Defence, the Home Office, the PSA and major health authorities are all aware of the gains to be had from the design and build approach. Likewise many of the major property developers have seen the advantages of the construction management system.

The 1993 survey was representative, capturing 12 per cent of the new construction orders during 1993. The authors of the 1991 report noted that the results may have overstated the use of the construction management approach. Six construction management projects were included, representing only 0.16 per cent of the total number of projects considered, yet their value constituted 19.36 per cent of the total value; one project alone accounted for £410m. The 1993 report is considered more realistic.

Whilst no similar surveys are known to exist in the UK civil engineering sector it is noted that the design and build approach is also becoming popular in this sector. As long ago as 1966 the Lofthouse Report,[3] which concentrated on road construction, identified the conventional separation of design and construction as the cause of elaborate designs and expensive construction.

In the 1970s contractor's design and build appeared in the UK civil sector with the adoption of contractor's alternative designs. The Kessock Bridge, Inverness for the Scottish Development Department, let in 1977, was one of the earliest and most significant major civil engineering design and build contracts.[4] Later other major design and build projects included the Docklands Light Rail, the Manchester Metrolink and the £270m. Second Severn Crossing. In 1991 the first design and build road contract was let by the Ministry of Transport and a year later in response to this trend the ICE introduced the standard design and build form of contract based on the ICE 6th.

Many of the traditional boundaries are beginning to be broken down; one notable example is in the tunnelling sector where some important UK promoters have decided to use the Institution of Chemical Engineers' Green Book (form of cost reimbursement contract conditions). The reason for this choice was that a payment mechanism closer to cost reimbursement rather than to a B of Q was seen to be appropriate for this highly uncertain work.[5]

A further recent development in the US construction industry is the concept of 'partnering'. This arrangement has been used by many major clients including Dupont, Procter & Gamble, Shell Oil and the US Government whilst Anglian Water have used a similar arrangement with contractor Biwater in the UK. The prime reason for introducing partnering is to enable the client to reduce his overheads by minimising his design and engineering facilities. The arrangement aims at promoting trust and co-operation between the parties rather than an adversarial relationship.

In its simplest terms partnering is a contractual arrangement between a client and his chosen contractor which is either open-ended or has a term of a given number of years rather than the duration of a specific project. During the life of the arrangement the contractor may be responsible for a number of projects, large or small, and continuing maintenance work and shut downs. The Biwater contract with Anglian Water ran for 3 years from 1989 and provided for work to be made available to Biwater each year – these were mainly design and construct projects on existing plants.

Partnering has other major benefits for both parties including enabling the introduction and continuing use of total quality management techniques, familiarisation with safety procedures and offers the opportunity for innovation. An excellent report on partnering has been published by NEDO.[6]

The traditional methods

The traditional method of contract procurement is based upon the rigid separation of design and construction. The promoter, usually after undertaking a feasibility study, appoints a team of consultants (led by architect/engineer) to undertake the detailed design. The design team prepares detailed drawings, specifications and often B of Q. The tender documents are prepared and the contract awarded, usually to the contractor with the lowest bid. The contractor manages the construction usually using subcontractors.

The traditional route is readily understood and is in general use and recommended by many practitioners, some of whom may have a vested interest in its use. However, it is generally becoming less popular as enlightened promoters seek more appropriate alternatives.

The strengths of the traditional approach are well documented and include:

- generally a high degree of certainty on the basis of the cost and specified performance before a commitment to build, however variations and claims can make this less so;
- clear accountability and tight control at every stage, again claims and variations can make this less so;
- competitive pricing between main contractors;
- opportunities to combine best design and contracting skills in well-understood relationships;
- allows for nomination of particular specialists by client;
- flexibility in developing the design up to the contract documentation stage, and if necessary varying the construction design, however the cost can become less certain;
- well tested, in practice and in law;
- opportunity for promoter to require contractor to design part of the works (through standard contract clauses or supplement).

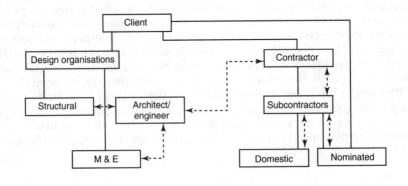

Figure 5.1 The traditional system

The weaknesses of the traditional approach include:

- the standard contract fosters an uneasy guarded relationship between the parties, which can easily become adversarial, instead of a collaborative relationship;
- the architect often nominates his preferred design team and is expected to control them yet has no contractual liability for their performance;
- the promoter is in direct contract with the contractor and yet has no right of communication or instruction under the contract;
- alternative methods allow commencement on site prior to completion of design and thus enable earlier completion of project;
- greater co-ordination and control is required because several parties with different contractual relationships are involved;
- split responsibilities – client is in contract with too many parties which can be a serious weakness in the event of major defects arising;
- there is no opportunity for the contractor to contribute his construction expertise in the design process.

If speed is a priority it is possible however to use *accelerated traditional methods*, usually through the use of 'two-stage tendering' or 'negotiated tendering' procedures. These enable design and construction to run more closely, securing some time saving but giving less certainty about cost. Chapter 7 'Tendering procedures and bid evaluation' includes a detailed section on the two-stage tendering approach including the examination of a case study.

The advantages of accelerated traditional methods are:

- two-stage tendering allows early testing of the market to establish price levels and gives early contractor involvement resulting in speed of construction;
- negotiated tendering allows early contractor involvement for 'fast tracking' (i.e. beginning work on site before the design is complete);
- negotiated tendering also gives flexibility for design development as the construction proceeds.

The possible disadvantages are:

- less certainty on price before a commitment to build;
- competition is reduced or absent in negotiated tendering and more approximate in two-stage tendering;
- more concentrated client involvement required to ensure efficient planning and control throughout the process.

The design and build methods

Back in 1964 the Banwell Report[7] remarked that 'in no other industry is the responsibility for design so far removed from the responsibility of construction'. Today the contractor-led design and build procurement option is the fastest growing sector of the construction market in the UK. Furthermore it is increasingly perceived as the appropriate answer for large and complex projects. Indeed some leading commentators now consider that design and build will become the norm by which other systems are judged. The method requires the contractor to take overall responsibility for both the design and the construction of the project. However, the client may appoint an independent quantity surveyor or a project manager to keep a check on quality and cost.

The design and build approach allows the contractor's design and construction team to consider, at the earliest conceptual stage, site-specific construction issues which a consultant working in isolation is not normally equipped to deal with. For example, on a large marine contract the team will be able to establish: if a site is suitable for the use of large cranes; whether heavy floating barges can be used in a tidal location; how materials will be transported to the construction locations; whether there are suitable areas available close to the site for setting up a precasting or pre-assembling yard; what skills are characteristic of the local labour force and how the local weather during the construction period will affect the construction methods.

The most economic type of structure and the most suitable method of construction will depend on the answers to the above questions, together with the contractor's specific expertise and the availability of construction equipment. It is at this stage when the combined team has at its disposal all the relevant facts and techniques that increased productivity may be considered – thus reducing the overall cost to the client.

The design and build approach is extremely flexible and many different versions have emerged over the past few years. The major difference between them is the amount of design input by the employer's designers and the contractor's designers.

The amount of tender documentation provided by the client (known in the JCT Standard Form of Building Contract with Contractor's Design 1981 (JCT 81) contract as the 'employer's requirements') can vary from little more than a written brief to a fully worked out scheme. The greater the priority the client gives to design the larger the amount of information tends to be included in the tender documents. If the client's priorities are economy and speed then less design information will be included, leaving more scope to the contractor.

The NEDO document *Thinking about Building*[8] identifies three main categories of the design and build approach:

- *direct*, where a designer/contractor is appointed after some appraisal but without any competition;
- *competitive*, where the conceptual design is prepared by consultants enabling several contractors to offer designs and prices in competition;
- *develop and construct*, where the employer's designers complete the design to a partial stage before asking contractors to complete and guarantee the design in competitive tender, either with their own designers or by taking on the employer's designers.

The strengths of the competitive design and build approach include:

- single point responsibility for the total design and construct process after the selection stage;

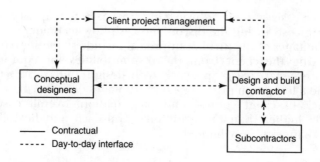

Figure 5.2   The design and build system

- competitive pricing – the overall cost to client may be the same or lower when compared to the traditional approach;[9]
- guaranteed cost of building and date for completion;
- can allow consultant design input;
- can speed up the commencement dates;
- input by contractor can lead to a more economical design;
- less adversarial than the traditional approach with fewer disputes.

The weaknesses of this approach are:

- competing schemes may not meet client's requirements for cost and performance, unless specified in detail before the bidding begins;
- client's clearly defined brief required at commencement;
- making changes after the start can prove expensive and result in major delays;
- analysis of tenders can become subjective;
- the promoter may be more involved in the details of the running of the project;
- substantial amendments are required to JCT 81 if the client wishes to secure the project in sections;
- under the provisions of JCT 81 the client may be able to impose changes only with the contractor's consent (clause 12.2).

Turnkey

The turnkey contract has been adopted for major construction works on multidisciplinary projects, particularly in the process plant sector, in the UK and overseas. Under turnkey contracts the entire process of design, specification, construction and commissioning is carried out by a contracting organisation often in joint ventures or consortia. Sometimes the client may wish the contractor to finance, operate and maintain the facility.

The client will normally issue a brief based on a performance specification together with outline drawings indicating a preferred layout. The contractor's lump sum bids are evaluated firstly on a technical and performance basis and secondly on a financial basis for capital expenditure and running costs (using the discounted cash-flow technique).

The advantages of the turnkey approach include single source responsibility relieving the client from the responsibilities for equipment and performance, a fast track approach with design and construction overlapping and a lump sum price. The disadvantages of the turnkey approach include lack of client control and participation, overall cost may be significantly higher than the traditional approach and flexibility to incorporate changes is very limited.

The management methods

Under these methods the contractor offers the client a consultant service based on a fee for co-ordinating, planning the construction, managing and executing the project. The approach ensures that the contractor (or construction management consultant) is part of the promoter's team from the outset ensuring that maximum construction experience is fed into the design.

Management contracting

In a management contract the permanent works are constructed under a series of construction contracts (also known as trades contracts or works contracts) placed by the management contractor after approval by the client. The services offered by the management contractor may include those given under the headings below.

Pre-contract
Programming the design, design input on 'buildability', budget and cost forecasts, advice on financing, cost control, materials procurement and expediting, preparation of tender documents, evaluation of tenders, selection of construction contractors, insurances and bonds – policy and implementation, construction planning and programming, methods of working.

Post-contract
Supervision and control of construction contractors, provision of central services and construction equipment, design of temporary works, quality control, industrial relations – policy and monitoring, costing of variations, certification of interim and final payments to construction contractors, assessment and monitoring of claims.

Amongst the reasons for a client selecting a management contracting approach could be:

- The project is large or complex.
- There is a need, or strong economic advantage, in an early start and early completion of the project in the situation where work was not sufficiently defined prior to construction. (This circumstance requires good planning and control of the design/construction overlap and careful packaging of construction contracts.)

- The need to consider particular construction methods during the design phase.
- In a complex project involving high technology this approach would provide greater flexibility for design change than conventional contracts.
- The project is organisationally complex requiring the management and co-ordination of a large number of contractors and contractual interfaces and several design organisations.
- The client and his advisers have insufficient in-house construction management resources for the project.
- The client wishes to select his own designers.

The strengths of management contracting can be summarised as follows:

- Time can be saved by a more extensive overlap of design and construction utilising the management contractor's expertise in construction planning.
- It allows flexibility particularly where the programme and design are ill-defined and subject to change.
- Cost savings can be achieved through better control of design changes, improved buildability, improved planning of design and construction into packages for phased tendering, keener prices due to increased competition on each package.
- It reduces delays and the knock-on effect of claims.
- It is easier to control the selection of construction contractors to those of known ability.
- It avoids adversarial attitudes – leads to a more harmonious relationship between the parties.

The weaknesses of the system can be summarised as:

- A client may be exposed to a greater risk due to: reliance on a contract cost plan prepared on the basis of incomplete information rather than a B of Q based on firm information, late information, works contractors failing to perform to time or quality standards (there is no direct contractual link between the client and the construction contractors), delays and subsequent time and cost overruns.

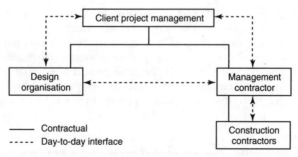

Figure 5.3 The management contracting system. *Source:* Hamilton 1993 Relationship Choices Part 2. *New Builder* 28 June 1993. Reproduced by permission of the Institution of Civil Engineers

- There is evidence that the overall construction cost may be greater under this fast track approach; however, this is normally offset by an early completion and the additional letting income or revenue accrued.
- There is a tendency to produce additional administration and some duplication of supervisory staff.
- The management contractor's ability to ensure compatibility between design and construction methods may be limited.
- The potential for interface 'grey areas' between works contractors is high.
- If the JCT management contract (JCT 87) is used in unamended format the client is responsible for the knock-on effects of non-performance of the works contractors, not the management contractor.

Construction management

In recent years in the UK we have witnessed the emergence of the North American construction management approach, particularly on major fast track property developments in London. In the USA the use of construction management in the building sector probably accounts for at least 40 per cent of the market; the majority of the remainder of projects are lump sum – bid on the basis of drawings and specifications. (A good overview of the American approach is contained in the CIOB publication *The Management of Construction Projects – Case Studies from the USA and UK*.[10])

The construction management system is similar to management contracting. However the construction management approach as practised in the UK does demand a client with the commitment and expertise to become involved in the development process as he/she employs the construction contractors direct. Under this approach the construction manager acts purely as the client's agent, in an impartial capacity co-ordinating and controlling all aspects of the project. Construction management is very much a team approach involving the client, designers and construction contractors, with the construction manager acting as the team leader.

It is recommended that the construction manager is appointed by the client at the same time as the architect/engineer, if not before. The consultant construction manager may be required to control the design process particularly where there are several design consultants. He would also provide expert advice at both the design and construction stages on construction planning, costs, construction techniques and buildability.

The strengths of construction management can be summarised as follows:

- As with the management contracting approach construction management offers a better chance of success with greater flexibility within a tight time-scale on a complex project.
- Construction management allows the client a full and continuous involvement in the project and a greater degree of control – indeed

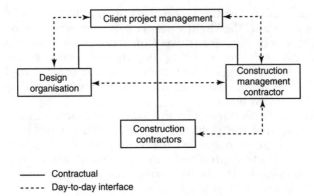

— Contractual
- - - - Day-to-day interface

Figure 5.4 The construction management system. *Source:* Hamilton 1993 Relationship Choices Part 2. *New Builder* 28 June 1993. Reproduced by permission of the Institution of Civil Engineers

the client is the ultimate decision maker between the designer and manager in order to effect the balance between architecture, technology, time and cost.

- The client has greater flexibility in the appointment of construction contractors, direct payment from the client to the construction contractors can result in lower bids and in theory there is a better long-standing relationship between the parties (however, see legal case of interest).
- It utilises a team concept thus avoiding traditional adversarial attitudes.

The weaknesses of the construction management approach can be identified thus:

- As the client contracts direct with the construction contractors the total risk in the event of failure or dispute lies with the client, there being no intermediary main contractor.
- Clients will have to contribute a great deal of expertise when undertaking construction management; this may prove to be a daunting experience for inexperienced users of the construction industry.
- The client does not know the overall tender price at commencement of the works.
- No standard form of contract is available (JCT form due in 1994).

UK case study of the construction management approach
The Broadgate development at Liverpool Street Station, London was very much the pioneer in the UK for the construction management system. Particularly impressive on this massive £500m. office block development was the speed of construction – the first two eight-storey blocks took only 12 months from the start of the foundations to the handover to the tenants.

The system demanded a different approach; out went the traditional terms of engagement for the designers, B of Q and the standard JCT form of contract and in came an integrated specially written set of contracts, innovative fast-track construction methods, changed attitudes and relationships and the client's permanent representative on site.

The construction manager (Bovis Construction on phases one to four) was responsible for co-ordinating the design, procurement and construction, for ensuring good buildability and for controlling budget and programme. Unlike the system in the USA they were not responsible for managing the design team.

The selection procedures for the trade contractors was rigorous in order to ensure that the right team was selected for the job. Only those contractors who the management team thought would fit in were invited to tender. The trade contractors were presented with a programme of work and a time-scale, they were further required to attend an interview to discuss their methods of working and introduce their key people who would be in charge.

Value engineering
The trade contractors had a significant contribution to make in the design process using the American technique of value engineering. In its simplest form value engineering can be described as a scientific method of analysing a product or service in order to reduce its overall cost. The approach differs from the traditional cost-cutting exercise in that it demands that neither the quality nor the performance of the end function be impaired in any way. By using value engineering techniques the function of the various components of the building process was analysed by the whole of the construction team and an alternative was sought to obtain the same end more cheaply.

Trade contractors could be involved at the very earliest stage of the design, providing them with a direct working relationship with the design consultant. In addition the procurement of the trade packages used 'scope documents' rather than detailed drawings and B of Q, allowing the contractor to be innovative – again to the client's advantage. A typical example of the benefit of value engineering was in the construction of the internal staircases. Initially these were to be steel – frames infilled on site with concrete. Following the evaluation process it was suggested that it would be cheaper and quicker and improve the quality if they were supplied with the concrete already in place.

Fast-track construction techniques
On the site many simple ideas were utilised in order to cut construction time including:

- buildings designed to a rigorous 7.5 × 7.5 m column grid ensuring economic steelwork design and repetition in erection;
- trade contractors brought in early in order that their expertise could be maximised, many were flown to the USA at the client's expense to learn more efficient techniques;
- apart from the concrete floors poured over metal decks all the wet trades were eliminated;
- basements limited to one level with the superstructure construction commencing as soon as foundations were complete;

- cranes lifting structural elements two or three at a time in chandelier fashion;
- off-site prefabrication of completely fitted out 10 tonne toilet pods and air-handling units;
- fire protection for the steelwork pumped from a central source;
- curtain walling preassembled into 7.5 × 3.5 m panels and delivered to site complete with glass.

Legal case of interest

Unfortunately not all the phases at Broadgate constructed using the construction management system have run smoothly. In 1990 the court of Appeal pronounced judgment in the case of *Rosehaugh Stanhope (Broadgate Phase 6) plc and Rosehaugh Stanhope (Broadgate Phase 7) plc* v. *Redpath Dorman Long Ltd* (1990).

Under the terms of the agreement between Rosehaugh Stanhope (the developer) and Redpath Dorman Long (the steel frame supplier) it was a condition of the contract that Redpath Dorman Long must complete work within the time allocated by the programme.

The contract further provided that in the event of delay resulting in the trade contractor failing to complete in time the construction manager was entitled to make a bona fide estimate not only of the loss that the employer had incurred but also any loss that he considered the employer might incur in the future. The estimate was expressed as being binding on both parties until final agreement. Furthermore the contract provided that the trade contractor should pay the amount of the construction manager's estimate to the employer.

Under Broadgate Phase 6 the amount of the construction manager's estimate of the delay was £5 122 418 with an amount of £5 741 730 estimated for Phase 7. Redpath Dorman Long claimed that they were entitled to an extension of time and appealed to the Court of Appeal after being issued with a writ by the developer. The Court of Appeal upheld the appeal taking the view that the trade contractors would have no obligation to pay the amount of the construction manager's estimate if they had an arguable case for being awarded an extension of time. Lord Justice Stoker said that his views accorded with commercial sense since otherwise Rosehaugh Stanhope would be in a position to enforce potentially ruinous payments by Redpath Dorman Long in circumstances in which it is later established they were under no obligation to pay.

The design and manage method

This approach combines some of the characteristics of design and build and management. It allows the client more opportunity to be involved in the design process and to make changes to the specification. The approach may be desirable on follow-on contracts, on fast-track contracts or where the contractor has some specialised expertise. The contractor is normally appointed early, often in competition based on a quoted fee with a build-up of the required preliminaries. All construction work is undertaken by specialist construction contractors.

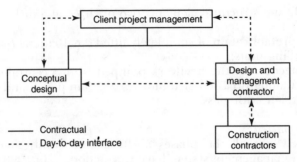

Figure 5.5 The design and management system. *Source:* Hamilton 1993 Relationship Choices Part 2. *New Builder* 28 June 1993. Reproduced by permission of the Institution of Civil Engineers

Two variants of design and manage are noted: 'Contractor' – in which the project design and management firm takes on the trade contractors and 'Consultant' – in which the project design and management firm acts as the client's agent with the trade contracts direct to the client.

Choosing the right strategy

Before identifying the important factors which should be considered prior to the selection of the appropriate contract strategy it is essential to consider those external matters which may affect the project. Typical external factors include:

1. *Political*: government policy – expenditure on roads, defence, housing, education, hospitals; self-funded infrastructure projects – BOT contracts; import and export regulations – over 50 per cent of the materials on some of the large London Docklands projects were imported; national and local planning laws and policy; new legislation, e.g. on contaminated sites, deregulation of professions; grants for developments.
2. *Economic*: interest rates, exchange rates, level of inflation, tax benefits, European Union requirements, market opportunities, hitting the market, e.g. shopping centre opening before Christmas, payback period, financial loopholes.
3. *Legal*: Acts of Parliament – over 100 government regulations can affect construction projects, availability of standard conditions of contract, contract law, case law, conciliation procedures prior to arbitration/ litigation, cost of dispute.
4. *Technological*: prefabrication, standardised buildings and products, innovative materials – limited suppliers, intelligent buildings, specialist's patent on design.

The priorities that should be identified by the client before selecting the appropriate strategy include those listed under the headings below:

Time

Speedy occupation of the new building, or the opening of the new bridge or transport system, may be more important than anything else. Inherently fast methods, like direct design and build or the management contracting, can best meet this objective.

Rapid completion can also be achieved with traditional methods either by compressing the time available for each stage of the project, or, more usually by overlapping stages. This means starting the later stages of design documentation or construction before earlier ones are complete, requiring greater management skill from all involved. Costs can be controlled by good management, but are not based at the outset on completed design information.

If speed is the main priority the client should allow for plenty of careful pre-planning of the work, for example to ensure that all the necessary legal approvals are obtained before the commencement of construction.

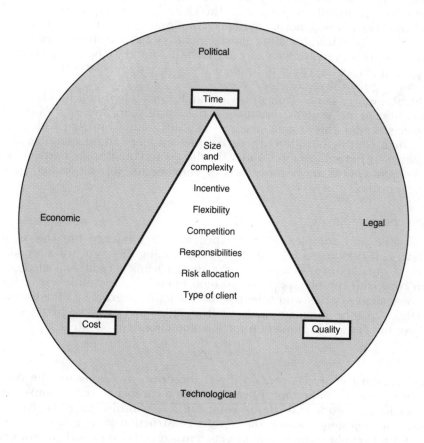

Figure 5.6   Factors affecting the choice of contract strategy

Cost

Some clients need to know the final contract figure before they can begin a project. Design and build paths can give a firm guaranteed price. Guaranteed maximum price contracts must include a premium cost for the additional risk carried by the contractor. Future maintenance costs should be considered during the design phase – extra initial cost may save money later.

The traditional sequential method will provide the clearest indication of the basis of price before construction; however, the final figure may be different due to costs incurred outside the contractor's control.

Price certainty before the contract is let is difficult to obtain if speed of completion is a priority, and near impossible if the works are varied to include late technological advances. Generally the quicker the project the greater the uncertainty of final price to the client.

Quality

The appearance and quality (functional, architectural, technical and workmanship) are usually important to the client. However, the best design and construction may not be available from one source. Previously it was thought that the traditional or management systems may have been more attractive. However, we have already seen that the develop and construct approach allows the client to appoint his own designers who then switch to working for the contractor.

Furthermore, James Pain in his research into design and build found that there is no evidence to suggest that quality, in any of its various respects, under this system need be any different from that with conventional contracts. Indeed there is some evidence to suggest that the quality may indeed be better under the design and build route particularly where the contractor is concerned to enhance his reputation and gain repeat orders.

Size and complexity

The scale and complexity of the project will influence the choice of strategy. If the project is complex and demanding the client will normally have a greater involvement with the design team. Traditional methods can cope with complex projects but tend to be high in claims; one of the management systems would be the most suitable alternative. Traditional methods may be adequate for large straightforward projects, particularly where the design is complete prior to commencement on site.

Incentive

Few standard contracts contain direct incentive provisions for the contractor to complete the contract earlier or for a cheaper price. If completion of sections of work and overall completion is critical a payment system linked to milestones within the agreed construction programme can be used. Cost reimbursement and target contracts can also provide incentives to the contractor for reducing costs.

Flexibility

If the client wishes to keep his options open in a fast-developing technology the management approach will generally provide the most flexibility. Variations generally cost more on the design and build paths than on the traditional paths where a basis for negotiating is available.

A further consideration is that the client may have good reasons for selecting specialist contractors or suppliers and the procurement route chosen should reflect this preference, for example by allowing nominated or named subcontractors or by permitting the client to employ such specialists direct.

Competition

Competitive tendering is normally required for all contracts; indeed the traditional approach puts all competitors on an equal basis.

Even though full information is not available for tendering with accelerated traditional contracts they can be competitive on the basis of typical rates for the sections of the work or on approximate measurements. Speed and full competition are compatible in management approaches because all work goes to competitive tender in planned packages with full documentation.

Unsuccessful tenderers under the competitive design and build approach may seek a fee to cover part of their costs because of the considerable work which has to be done to prepare such a tender.

Responsibilities

The various procurement paths each give different relationships and patterns of responsibility to the participants. Clients should consider which path gives them the most acceptable pattern.

With the traditional path, the design team and contractor are separately and directly responsible to the client, but responsibility for the whole project is divided between them.

Design and build paths unify the responsibility for the whole project, but tend to place the designer at a distance from the client. The designer will be responsible to the design and build contractor and not to the client should any dispute arise. It is wise therefore for the client to employ a design team, including a quantity surveyor, to look after his interests and who will be directly responsible to him.

Management systems divide responsibility into three layers: the design team; the construction manager; and the trade contractors who carry out the work. However, because the management element can be set up at the start and run right through the project, few problems are likely to arise from the transfer of information between designers and contractors.

Risk allocation

Undertaking any project, however well planned, involves the client in taking the risk that things may not go exactly as expected. On some procurement paths the client can share some or most of this risk with a contractor for a cost; on others the client will retain it. The client may wish to retain the risk in order to keep control of the project, or to be able to intervene to update the design.

Even though design teams agree to act skilfully it is not their function to guarantee a time for completion or the final cost. Contractors on the other hand, sign contracts which guarantee project cost and timing. Agreed compensation can be levied to recover losses if they do not perform. On the design and build routes the contractor accepts the majority of the risk.

Traditional paths divide the risk between the client and the contractor. On management paths most of the risk remains with the client, whether it be management contracting or construction management.

Type of client

Previously public clients in the UK usually adopted the traditional approach as the normal contract strategy. Now with the privatisation of much of the public sector a far more enlightened approach is prevalent.

Regional health authorities are also far more enlightened and are beginning to select the design and build approach for major hospitals. Reference 11 describes one of the first design and build contracts for a major UK hospital. The £51m. district general hospital in Dunfermline, Scotland was completed within 46 months from approval – slashing, it is claimed, up to 6 years off the traditional gestation period for this size of hospital.

Many local authorities, however, are still slow to change and keep to the tried and tested traditional methods whether or not these are appropriate for the circumstances. Arbitration or litigation can, however, often indicate that the wrong contract strategy was chosen.

Project manager or client's representative

As construction works become more complex and are required in a shorter time many clients, particularly those who do not possess the necessary in-house expertise, are turning to a project management (PM) approach.

Under this system a single individual is appointed for a fee as the client's representative or project manager. It is recommended that the individual, who in practice may head a small team, has expertise in design, management and construction relevant to the particular project.

The project manager, who acts as the leader in an independent capacity, should be appointed at the earliest opportunity and should be involved in all stages of the project cycle from appraisal through to commissioning. Standard items of engagement of project managers have now been produced by the RICS.

One of the most important duties of the project manager should be to recommend the appropriate contract strategy. For most projects this should be done at the *concept design stage* via a workshop approach with the key personnel, i.e. the client, designers and project manager, present. This approach encourages teamwork and leads to the solution being owned by all members of the team.

The workshop approach requires the members of the team to identify and agree a list of critical characteristics for the project, to determine the relative importance of each and to rank them in order of merit. The team then analyses each procurement route by applying a weighting, or 'utility factor', to each of the client's objectives according to how well that route meets the objective. The most appropriate procurement route is then selected after totalling the product of each of the client's weighted objectives multiplied by each utility factor for each organisational method, the recommended contract strategy being the one that achieves the highest mark. For further reading on this subject see Skitmore and Marsden (1988), Bennett and Grice (1990) and Gillespie (1994).

Tutorial questions

1. The developer of the science park requires large-scale demolition and groundworks (approximately £0.5m.) to be executed in advance of other building works in order that the site is made more attractive for marketing purposes and will also enable a separate contractor to commence work on part of the site in one month's time.

   Identify the options available to the developer in order to secure his objectives, discuss the associated risks with these approaches and prescribe what form the contract would take. (RICS Final (1987 syllabus), specimen paper)

2. Your client, a large multinational organisation, wishes to erect a new headquarters in London. It is an innovative and dynamic company and it wants these qualities to be reflected in its new building.

   The organisation has appointed an architect well known for his revolutionary designs using the latest technology. You have previously worked with the architect and know him to be exacting in matters of quality.

   Much design development work will be carried out on a co-operative basis between the architect and the specialist manufacturers. The client requires that the building is erected in the shortest possible time and for a start to be made on site within 3 months of your appointment.
   Describe:
   (i) the contractual arrangements you would advise, identifying the relationship between, and the responsibilities of, the various persons who would be involved with the project;
   (ii) how control of, and responsibility for, design, time and quality would be achieved;
   (iii) how you would ensure that control of cost was achieved;
   (iv) the special qualities you would expect the contractor to have

   (RICS Direct Membership Examination, specimen paper)

3. With high interest rates on borrowed capital and a high proportion of construction cost related to timing of work, timely completion is vital to client's and contractor's financial control

   Discuss the statement: 'Most contracts do not give enough emphasis to time considerations.'

4. (a) Describe the procedure leading to the selection of the construction manager.
    (b) Identify the factors which should be considered by the construction manager prior to recommending the appointment of a trade contractor.
    (c) What advice should construction professionals give to their clients who wish to use the construction management approach?
    (d) What practical steps can be taken by construction professionals wishing to offer a construction management service to their clients?
5. Using the chart on pp. 6 and 7 of the NEDO document *Thinking about Building*,[8] identify a possible appropriate procurement path for the following projects:
    (a) 300 bed hotel in central London for an international hotel group;
    (b) 15,000 seater stand for Wolverhampton Wanderers FC;
    (c) hospital maternity block at the district general hospital;
    (d) headquarters for an international computing company

References

1. Yates A 1991 Procurement and construction management. In Venmore-Rowland P, Brandon P, Mole T (eds). *Investment, Procurement and Performance in Construction* E & F N Spon
2. Davis, Langdon, Everest 1994 *Contracts in use: A Survey of Building Contracts in use during 1993* RICS.
3. NEDO and EDC for Civil Engineering (Lofthouse Report) 1966 Efficiency in road construction, HMSO, London
4. Clements L 1984 The Kessock Bridge design-and-build contract, and proposals for managing similar contracts. *Proc. Instn Civi. Engrs* Part 1, **76** (February); 23–34
5. Fawcett D 1991 Debate: IChem.E versus ICE. *Tunnels and Tunnelling* June: 68–9
6. NEDO 1991 *Partnering: Contracting without Conflict*
7. Banwell H 1964 *The Placing and Management Contracts for Building and Civil Engineering* HMSO
8. NEDO 1985 *Thinking about Building*
9. Pain J, Bennett J 1988 JCT with Contractor's Design form of contract: a study in use. *Construction Management and Economics* **6**: 307–37.
10. Nahapiet J, Nahapiet H 1985 *The Management of Construction Projects – Case Studies from the USA and UK* CIOB
11. Hayward D 1993 Smooth operation. *New Builder* 12 March: 12, 13

Further Reading

Allen J D 1986 Construction management keeps Broadgate on fast track. *Construction News* 3 April, 22–25

Archer F H 1985 Executive Project Director, John Lok/Wimpey Joint Venture, Headquarters for the Hong Kong and Shanghai Banking Corporation, correspondence with the author, 3 April

Benaim R 1987 Teamwork key to design/build viability. Letters, *New Civil Engineer*, 15 January: 32

Bennett J 1986 *Construction Management and the Chartered Quantity Surveyor* Surveyors' Publications, March

Bennett J, Grice A 1990 In Brandon P S (ed) *Quantity Surveying Techniques: New Directions* Blackwell Scientific pp. 243–262.

Construction Management Forum Report and Guidance 1991 Centre for Strategic Studies in Construction, University of Reading

Dawson H 1988 Design & build: a client's viewpoint. *Chartered Quantity Surveyor* November: 29

Franks J 1992 *Building Procurement Systems* 2nd edn CIOB

Gillespie B 1994 Procurement route. *Building* 29 July: 46

Hancock M 1987 The selection of contractual arrangements. *Chartered Quantity Surveyor* July 12–13

Huntley C 1987 The future according to Broadgate. *Building* 8 May: 56–7

Janssens D E L 1991 *Design–Build Explained* Macmillan

Knowles R 1990 Onerous conditions – is the end in sight? *Chartered Quantity Surveyor* August, 21

Kweku K, Bentil M S, Herbsman Z 1987 Construction Management – is it really the way to go? *Managing Construction Worldwide*, Vol 1, *Systems for Managing Construction* Spon/CIOB/CIB, pp. 4–15

Macpherson I 1987 Faster and faster track. *Building* 15 May: 64–5

Macpherson I 1987 Can concrete steal the show? *Building* 22 May: 80–1

Macpherson I 1987 Real value for money. *Building* 5 June: 62–3

Masterman J W E 1992 *An Introduction to Building Procurement Systems* E & F N Spon

Morledge R 1987 The effective choice of building procurement method. *Chartered Quantity Surveyor* July: 26

Morris M R 1986 Construction management – inspiration for QS diversification. *Chartered Quantity Surveyor* November: 23–5

NEDO 1988 *Faster Building for Commerce*

Pain J 1990 Design-and-build contracts, what are they and do they work? *Estates Gazette* 6 October: 20–2

Pickup D P 1986 Alternative methods of health building procurement (unpublished paper)

Reading Construction Management Forum – forum for controversy. *Contract Journal* 17 January 1991: 11–13

Ridout G 1989 Construction Management Forum – Reading lessons. *Building* 4 August: 46–50

Skitmore R M, Marsden D E 1988 Which procurement system? Towards a universal procurement selection technique. *Construction Management and Economics* **6:** 71–89

Turner A 1988 Thinking about procurement. *Chartered Quantity Surveyor* November: 16

Turner A 1990 *Building Procurement* Macmillan

6

Type of contract

Introduction

The type of contract can be categorised by the type of payment systems which are either:

1. Price based – lump sum or remeasurement with prices being submitted by the contractor in his bid, or
2. Cost based – cost reimbursable and target cost. The actual costs incurred by the contractor are reimbursed together with a fee to cover overheads and profit.

Successive government reports[1, 2] have confirmed that the traditional unit-priced contracts (i.e. based on a bill of quantities (B of Q)) which are used on most civil engineering projects are too inflexible to deal adequately with the diverse requirements of both employer and contractor. Many of these projects, particularly the major ones, have been subject to large time overruns, extensive variations and substantial increases from the tender price to the final price.

Furthermore, we have already identified that the traditional route cannot easily accommodate excessive changes in design or overlaps in design and construction. Additionally the contractor is usually excluded from the design of the permanent works and the process often generates an excessively adversarial relationship.

It is therefore concluded that traditional contracts should not be awarded unless the project scope, including the detailed design, is 'sufficiently known'.[3] Many leading commentators also now believe that a greater degree of co-operation and trust in contractual arrangements generally, can result in a decrease in the end cost of a project.[4] As a result different strategies have been developed which are particularly relevant when considering major projects.

Lump sum contracts

The lump sum method is generally used on smaller straightforward works in the UK (referred to as 'plan and specification' contracts). However, this approach is used on some large projects and is frequently encountered in the USA.

Under the lump sum system each contractor is required to estimate the quantities and value the works, based on the client's designer's drawings and specification, using whatever methods considered suitable. Contractors could then either be required to submit one lump sum for the whole works, or to give a breakdown of the total sum against major activities or sections of the work. Payments are usually on a monthly basis and can be linked to the achievement of progress milestones.

In the UK building sector a lump sum approach is found in entire contracts. Here the contractor prices the B of Q and offers to build the entire project for the total sum tendered (e.g. under the JCT 80 (Private with Quantities)). The tendered sum thus quoted by the contractor would normally be subject to adjustment only for fluctuations, variations and claims.

In contrast in the USA, in some instances, contractors may have only a few days in which to prepare their bids and thus of necessity base it on a much simpler calculation than their UK counterparts.

Lump sum contracts demand that the design is complete at tender stage with minimal variations expected. In the USA variations on lump sum contracts generally require individual lump sum cost estimates and approval before the work is executed. Such a procedure is not so rigidly adhered to in the UK and the price for variations is often negotiated between the parties after the work is complete.

Some lump sum contracts include typical 'schedules of rates' (with or without quantities) requiring the contractor to insert individual rates for the sole purpose of valuing variations. However, in practice these rates may prove to be inappropriate for valuing the actual variations.

Admeasurement contracts

In the UK the most popular form of contracting in both the building and civil engineering sectors is still the traditional system. The project is designed by the client's consultant designer (architect or engineer) and detailed B of Q are prepared in accordance with appropriate standard methods of measurement (unit price method).

Most civil engineering contracts, which often contain a substantial amount of unpredictable work below ground, are admeasure contracts and contain provisions for remeasurement on completion (e.g. under the ICE 6th). This system clearly minimises the contractor's risk as he will obtain reimbursement for the quantities executed measured in accordance with the standard method of measurement rules.

For some considerable time it had been assumed, in the UK, that B of Q (the unit price method) offered advantages when compared to the lump sum approach. However, in recent years there has been a shift away from the traditional approach by employers, particularly on major construction works with many contracts placed using versions of management contracting/construction management, contractor's design–build and design–manage.

It is apparent that there is a considerable amount of vested interest in maintaining the status quo as far as measurement and B of Q are concerned. It would therefore be useful to consider independently the procedures under B of Q in some detail in order to identify the perceived strengths and weaknesses of the system:

1. Prompts the client and design team to finalise the design before the bill can be prepared.
It is claimed that the discipline imposed on the design consultants in providing the quantity surveyor with the information required for the B of Q preparation ensures that the design is fully thought through at pre-tender stage, avoiding unnecessary and expensive post-contract variations.

This is obviously true on many projects; however, on others B of Q are prepared from partially complete design drawings with the resultant variations after the contract is let. Indeed the B of Q may help to conceal the absence of sufficient pre-planning or investigation.

Quantities should be an accurate reflection of the work, however it is not unusual for them to be overestimated if the design is incomplete. This may expose the promoter to an additional risk following remeasurement. For example, under the provisions of clause 56(2) of the ICE 6th the contractor could be entitled to request a renegotiation of rates 'should the actual quantities in respect of any item be greater or lesser than those stated in the Bill of Quantities'. Thus essentially the promoter guarantees the accuracy of the quantities within the B of Q and undertakes to compensate the contractor in the event of any discrepancy.

It is perhaps worth stating that the economics of contracts generally indicate that any increases in quantity usually justify a reduction in rate whereas lesser quantities justify an increase. However, in practice contractors often seem to be able to justify an increase in rate whether the quantities are lesser or greater!

Furthermore it is worth reiterating the comments of Professor Peter Thompson of UMIST: he considers that if the work is not fully defined then the contracts should not be placed on a 'unit price' basis. Indeed the fact that there are so many claims under these types of contract indicates that the contract strategy and form of contract are incorrect.

2. Avoids the need for all the contractors to measure the works themselves before bidding, and avoids duplication of effort with resultant increase in contractor's overheads which are eventually passed on to clients.
It is acknowledged that B of Q do give the tenderers a clear idea of the scope of the work involved. However, B of Q based on standard methods of measurement, particularly in the building sector, are extremely detailed. Indeed a frequently made observation is that 80 per cent of the cost is covered in 20 per cent of the items. This level of detail merely helps to increase: the time taken to prepare, the technical skill level required to measure, and the quantity surveyor's fee for preparation of, the bills. Furthermore, historically few B of Q have dealt satisfactorily with the

M&E content, leaving typically 30–40 per cent of the value of building projects as 'prime cost' lump sums.

In contrast, contractors in the USA manage to price construction works without the use of B of Q. Indeed Carr[5] considers that the US system where contractors prepare their own quick quantities is in fact more efficient and cheaper overall than the traditional UK system.

3. Provides a commonality in tenders thus providing the opportunity for realistic tender evaluation.
If the priced B of Q are submitted with the tenders the contract administrator obviously has a vast amount of pricing information which can be used for comparison of the bids.

However, contractors rarely bid on the same basis, some may 'front-end' load rates for work required to be executed early (e.g. excavation) in order to secure additional revenue to finance the project. This practice shifts the balance of risk to the promoter, particularly in the case of a contractor's serious default or bankruptcy after completion of the early works. There is thus uncertainty as to whether the lowest tenderers will yield the lowest final price.

Furthermore, a contractor's estimators may spend some time attempting to discover errors or underestimates in the original quantities and then load or lighten individual rates with the intention of securing a substantial source of additional profit.

The client's contract administrator can of course compare individual unit rates between all tendering contractors in order to identify if there are any 'rogue' rates in the lowest tenderer's bid. He could then attempt to negotiate with the lowest tenderer to adjust individual B of Q rates within the same tendered total.

However, renegotiation is not recommended by the Codes of Practice as it introduces a second-stage tendering approach which is obviously unfair to the other contractors and may result in an unsatisfactory 'horse-deal'.

Under the B of Q system the contractor is generally paid for the amount of work he actually does. However, it has been demonstrated that contractors can increase their profit by manipulation of the rates within the B of Q to the obvious disadvantage of the promoter. The contract administrator should be alert to such ploys and may have to recommend that the lowest bid is too high a risk for the promoter and should be rejected; a far from easy course of action on local authority or public works contracts.

4. The unique coding system identified in the method of measurement against each item enables contractors to utilise computers efficiently for estimating B of Q.
Again true in theory but not so easy in practice. There are many different standard methods of measurement all with different rules for measuring the same construction activity, e.g. SMM6, SMM7, CESMM2, CESMM3, the Highways Method of Measurement, the International Method of Measurement and the unique methods of measurement of some large authorities.

Furthermore, the same B of Q descriptions may be required to be priced differently due to different cost considerations identified on the drawings, the soil reports, the specifications or due to conditions on site. For example, one basement excavation may be in firm ground enabling a neat volume to be excavated with steeply battered sides. Another basement may be in poor ground requiring extensive dewatering and continuous sheet piling – none of which is measured in the bills and should be included by the contractor in his rates. Both basements could be described and measured similarly in the B of Q.

5. Can be used as a basis for monthly interim valuations.
Agreed, but interim valuations can easily become far too detailed and take too long to prepare. It is not unusual on a major project for both parties to spend at least a week on the preparation of the interim valuation.

There are much easier and quicker ways of assessing a realistic value of the monthly interim value of construction works – e.g. based on percentages of lump sums quoted against major activities or based on a typical 'S' curve as the UK government contract GC/Works/1 (Edition 3) (Fig. 6.1).

Interim payments under lump sum contracts can be linked to completed stages of construction, thus benefiting the promoter by encouraging the contractor to maintain adequate progress. On the other hand, contracts using B of Q rarely contain such incentives to maintain progress; generally the contractor is entitled to payment in full for the work done (less retention) even if the progress is unsatisfactory.

Incentives are particularly important on major projects in order to ensure that contractors meet their obligations. The incentive could take the form of a significant payment after the timely completion of a certain stage of the work, e.g. design. A further incentive, used by the Central Electricity Generating Board on Heysham 2 nuclear power station, was to use a 'key date' system of 6-monthly reviews which impinged directly on the cash flow of the contractor if progress was not maintained.[6]

6. Rates contained in bills can be used as a basis for the valuation of variations.
True they can and the rules for the valuation of variations contained within the JCT 80 and the ICE 6th positively require this as a starting-point – all well and good for small straightforward variations.

However, the client's representative often needs to know the detailed build-up of the contractor's rates before he can calculate even the simplest pro rata rate. Furthermore there will be many occasions, however, when the rates in the B of Q, or a calculation based on those rates, may not be appropriate.

Under some circumstances basing the value of variations on B of Q rates will be positively unreasonable. Often on contracts under the JCT 80 and the ICE 6th the parties may wait until after the changed work is complete before attempting to value the variation. The contract administrator may then be looking to impose B of Q rates or pro rata thereto whilst the contractor's commercial manager will be looking to recover cost on a 'cost reimbursement' or 'daywork' basis – with every possibility of a dispute developing between the parties.

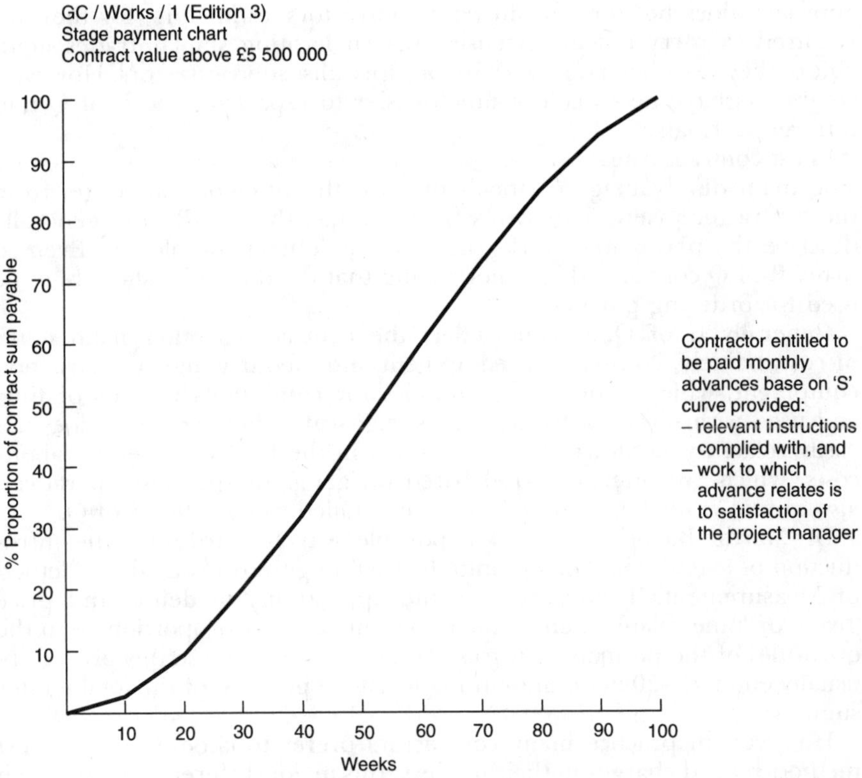

GC / Works / 1 (Edition 3)
Stage payment chart
Contract value above £5 500 000

Contractor entitled to
be paid monthly
advances base on 'S'
curve provided:
– relevant instructions
  complied with, and
– work to which
  advance relates is
  to satisfaction of
  the project manager

Figure 6.1   GC/Works 1 (Edition 3) – standard 'S' curve included in contract for
payments

The government contract GC/Works1 (Edition 3) includes a positive
provision relating to the valuation of variations. It provides that interim
payments for variations are paid in full provided that the value of the
variation is agreed in advance. In contrast if the parties cannot agree in
advance then only 95 per cent of the quantity surveyor's value based on
the traditional method using the B of Q rates will be included in interim
payments.

Chapter 11 considers in detail the principles of, and the problems
associated with, valuing variations particularly when B of Q are used.

7. Can assist the parties in the control and financial management of the works.
(a) Pre-contract estimating.
It is true that the client's cost adviser can use data from previous projects
when establishing the initial budget price for a similar new project.
However, in the case of the RICS Building Cost Information Service
(BCIS) it is necessary for all B of Q compilers to separate the documentation into the standard functional elements.

Unless the B of Q are repackaged into the conventional format this approach does not find favour with contractors' estimators, for they are required to carry out an extensive search locating similar trades items before they can obtain quotations from specialist subcontractors. However, computers have now made it much easier to repackage the B of Q into different formats.

(b) Post-contract control.

The main disadvantage of the B of Q at the post-contract stage, from the contractor's viewpoint, stems from the fact that the B of Q generally describe the permanent works or materials left on completion. Even so many B of Q contain a disclaimer stating that the quantities should not be used for ordering purposes.

Generally B of Q, do not reflect the contractor's other main items of cost, namely, construction equipment and labour gangs. Construction equipment, which is often kept on site for continuous periods of time and used with several activities/trades, e.g. scaffolding, cranage, does not easily relate to the descriptions contained in the B of Q. Likewise labour costs, which are often incurred based on gangs of specialist operatives, also cannot easily be separated into the detailed rates in the B of Q.

Dr Martin Barnes identified a possible way forward with the introduction of *method-related charges* into the Civil Engineering Standard Method of Measurement. Tenderers have the opportunity to define and price 'fixed' or 'time related' items which normally are not proportionate to the quantities of the permanent works. Tenderers who choose this procedure usually enter 10–20 items amounting to 20–30 per cent of the total tender sum.[7]

However, in practice many contractors prefer to choose not to insert method-related charges in their tenders; this major difference in approach in contractors' tendering strategies can cause difficulties when comparing tenders.

(c) Development of contractor's cost management system.

Contractors have attempted to use B of Q for financial control on projects with some success on straightforward projects; unfortunately, however, major projects are rarely straightforward. For example, it is possible to prepare a crude financial monitoring system based on the anticipated cash flow for the project utilising the construction programme together with the priced B of Q. The actual monthly revenue received from the client can then be compared with the return anticipated at tender stage and any shortfall identified. In practice, however, the system can easily become less than useful particularly where large-value materials are delivered late.

Furthermore, some contractors attempt financial control by analysing, on award of the contract, the whole of the individual B of Q rates into labour, plant, materials profit and overheads – a mammoth task in itself on a major project.

Each month thereafter the revenue from the interim valuation can be compared against the actual cost incurred for each element of work in order to identify major discrepancies. Unfortunately on completion of such a review it is often too late to do anything about any adverse situation. (Fundamental law of project management – you cannot control the past, only the future.) Furthermore, a contractor may kid himself into believing that a computer can help in this process but may end up in creating a monster that needs to be fed large amounts of detailed data each month.

In other words, the whole system of B of Q may in fact divert contractors from designing an effective post-contract control system.

Conclusion

In conclusion it has been demonstrated that the use of B of Q may lull promoters into a false sense of financial security and may not in fact be suitable for some major construction works, particularly if the design is not complete.

However, when the design is complete the traditional system with B of Q has been found to be a perfectly adequate mechanism for selecting contractors and assisting in the financial administration of contracts. In the 10-year period 1974–84 the HKMTR after using various alternatives ended up using full B of Q based on consultants' detailed design – the only modification to the traditional system being that the payments were linked to the achievement of milestones.

As B of Q are so prevalent within the UK most contractors and subcontractors are familiar with their application. Indeed it is observed that some form of B of Q are used with alternative contract strategies, for example in the management method as a basis for the works contracts or internally by contractors when bidding for design–build projects. It seems that the UK construction industry is reluctant to give up the beloved B of Q!

Computer-aided design may, in the future, perform some of the work for B of Q production, but traditional measurement will still be required for the bulk of the valuation of work at post-tender stage.

A current SERC research project at Dundee University, due for completion in late 1994, recommends simplifying the B of Q to 20 per cent of its existing scope with a greater emphasis on fixed charges and time-related charges which in turn are linked to the construction programme. The outcome of this research into a 'resource significance cost and time model' is awaited with interest.

Finally, it is interesting to note that 'option A' in the New Engineering Contract provides for a 'Priced contract with activity schedule' – again a system which allows the potential for the creation of a computerised project cost model in which major activities and construction methods can be identified together with their associated costs and time.

Cost reimbursable contracts

Under cost reimbursement contracts the contractor is paid a fee for overheads and profit based on the actual cost of construction. The contractor's accounts are open to the client, with payments made monthly. These type of contracts can provide a valid alternative to the traditional approach in certain circumstances:

- where the risk analysis has shown that the risks are unconventional in nature or magnitude;

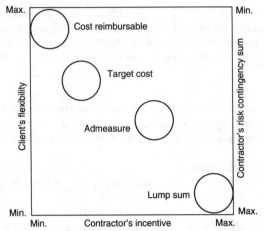

Figure 6.2   Relationships between types of contracts. *Source*: Ridout[8]. Reproduced
by permission of Reed Business Publishing

- where the engineer is unable to define clearly the works at tender
  stage, substantial variations are anticipated, or there is an emphasis
  on early completion;
- where an increased involvement of the employer and/or contractor is
  required or desirable;
- where there exists exceptional complexity, e.g. in multi-contract projects
  or where a high degree of technical innovation is demanded;
- emergency situations where 'time is of the essence';
- where new technologies or techniques are involved;
- where there is already an excellent relationship between client and
  constructor, i.e. there is trust.

Generally cost reimbursement contracts eliminate a large number of risks
in the project for the contractor and place them with the employer
who effectively becomes the project manager. It is recognised that the
employer gains greater flexibility in a cost reimbursement contract and
the contractor can be confident of an equitable payment for changed and
unforeseen events.

Cost reimbursement contracts allow contractors early involvement at the
design stage and allow clients participation in the contract management.
This approach identifies with the philosophy that it is cheaper in the long
run for the employer to pay for what does happen rather than for what
the contractor thought might happen in those areas of doubt which the
contractor cannot influence.

Target cost contracts

Incentives

Very few construction contracts within the UK include positive incentives
for performance, most rely on damages for non-performance e.g. liquidated
damages, maintenance periods, retentions, bonds and warranties. In con-

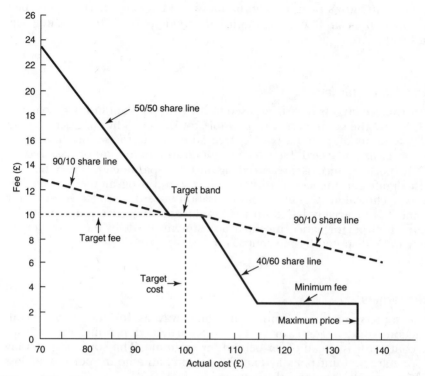

Figure 6.3 Target cost contracts – apportionment of costs. *Source*: CIRIA. Report 85, Target and Cost-Reimbursable Construction Contracts, 1982

trast it was estimated that in 1984 some 12 per cent of construction contracts carried out in the USA had some form of positive incentive.[9]

However, greater interest has been shown in positive incentives in the UK with the success of the so-called 'lane rental contracts' for motorway reconstruction. Under these contracts, a contractor is offered a bonus to complete work early, but is charged for each day it overruns the contract period. Contractors have consistently demonstrated that they are able to improve on the Ministry of Transport minimum performance standards.

Target cost contracts require a different approach when compared to traditional contracts. Target contracts demand that the promoter, contractor and the engineer are all involved in the management and joint planning of the contract. The promoter furthermore is involved directly in the costing and influencing decisions made on risk.

Target contracts have two main characteristics: they are cost reimbursable contracts and positive incentives are employed. A target contract is thus a means of sharing the risk between the employer and contractor with the latter being encouraged to maximise his performance. Similarly, over-expenditure is usually apportioned so that the contractor suffers most by receiving a reduction in the fee covering head office costs and profit.

The contractor's progress can be monitored against three targets, performance, time and cost, although combinations of different categories can be applied.

Performance targets

Performance targets have been used in the process plant industry, particularly where the contractor is responsible for design. The contractor either earns a bonus or incurs a penalty that adds or subtracts from his earned fee, or earns an 'award' fee which is added to a minimum or base fee.

The performance is measured against the parameter which has the most significant impact on construction cost and programme, e.g. quality, safety, technical management and utilisation of resources. However, as the payment is invariably based on the employer's subjective judgement of a contractor's performance, incentive adjustments made by the employer are likely to be disputed by the contractor.

Time targets

Time targets operate in much the same way as for motorway reconstruction work; a bonus or penalty is awarded, depending on whether the contractor is ahead of, or behind, programme. The bonus is generally a monetary amount per day and the penalty, an amount per day or loss of fee on work done past the completion date.

Cost targets

A typical cost incentive system involves the sharing of target project cost overrun and underrun. Usually cost targets are set for the combined construction costs with an agreed separate fee for overheads and profit. Other cost incentives which have been used more commonly in the process plant industry, involve the sharing of target man-hour cost overrun and underrun, or involve a bonus/penalty system based on the average man-hour cost.

Targets should not be fixed until the design is 40–60 per cent complete. The most common method of target setting appears to involve the use of a crude priced B of Q (reflecting the major cost-significant items – 80 per cent cost in 20 per cent of the items). However, in order for the incentive to be maintained the target cost must be adjusted for changes in the scope of the work, major variations and inflation.

Many target cost contracts, based on the guidelines contained in CIRIA Report 85, have now been successfully completed. Ridout[8] describes in detail two such contracts: Lerwick B Power Station in the north of Scotland and the 312 km long road project between Wino and Makambako in Tanzania.

Tutorial questions

1. On a multi-million pound interdisciplinary project based on B of Q, the client has suggested that, to ease the task of interim valuations, payment should be made based on overall percentage progress against the construction programme. Discuss the implications, advantages and/or disadvantages of this suggestion. (RICS, Direct Membership Examination, 1985, Project Cost Management paper)

2. 'The use of bills of quantities as a contract document may be considered outmoded practice.' Discuss this point of view. (CIOB Member Part II, Contract Administration, 1987)

3. (a) What are method-related charges? Discuss their use and benefits giving examples. How are method-related charges included in a contract under the ICE 6th?

   (b) A contract includes a method-related charge of £1000 per week for 50 weeks for a concrete batching plant during a 2-year contract. The batching plant remains on site for 54 weeks rather than the programmed 50 weeks. When and what should the engineer consider when the contractor asks him to continue to certify payment for the plant at £1000 per week?

   (ICE, Examination in Civil Engineering Law and Contract Procedure, 1992, Paper 2)

4. 'The ICE Conditions of Contract are drafted as an admeasurement type of contract'. Explain what is meant by this statement and give a detailed account of how this operates. Discuss the advantages and disadvantages of such a type of contract.

   If the contractor has a very high rate against a quantity that increases fivefold what action should the engineer take in accordance with the ICE 6th? (ICE, Examination in Civil Engineering Law and Contract Procedure, 1992, Paper 2)

5. If it is considered, in preparing a B of Q, that provision should be made for contingencies, how should this be done in accordance with CESMM3?

   Three contractors prepared tenders for a project based on the ICE 6th and CESMM3, for which the contract period was 40 weeks.

   Tenderer A priced the bill at a net cost of £75 000, added his overheads and profit at £25 000 and as a lump sum addition submitted a tender price of £100 000 and offered to complete in 36 weeks.

   Tenderer B examined the documents carefully and checked the quantities. He came to the conclusion that the bill was overmeasured by 15 per cent in value and that the net cost of executing the work required was £65 217. He constructed his tender thus:

Overhead charges and profit	£25 000
Net value of all measured work	£55 000
Method-related charges	£20 000
Total	£100 000

He added a lump sum addition of £25 000 for overheads and profit. When finalising his bid, he reduced the method-related charges by £2000 and submitted a tender price of £98 000.

Tenderer C also checked the quantities and came to the same conclusion as B. He submitted a conforming tender of £100 000 on the same basis as A, but offered as an alternative to carry out the work for a lump sum of £96 000.

How would you advise the employer on receipt of these tenders? In each of the three cases, what would be the value of the final account (assuming B and C's conclusions to be correct) and what profit would each make? (ICE, Examination in Civil Engineering Law and Contract Procedure, 1992, Paper 3)

References

1. NEDO 1979 *What's wrong on site?*
2. NEDO 1982 *Guidelines for the Management of Major Proposals in the Process Industries*
3. CIRIA 1982 *Target and Cost Reimbursable Construction Contracts* CIRIA Report 85: Part A – A study of their use and implications, Part B – Management and financial implications, Part C – Preparation of contract conditions
4. Millwood M 1983 Changing methods of placing contracts. *Chartered Quantity Surveyor* December: 177
5. Carr R I 1987 In Lansley P R, Harlow P A (eds) *Contrast of British and American Cost Practices, Managing Construction Worldwide*, Vol One, *Systems for Managing Construction*, CIOB/CIB, E & F N Spon pp 296–303
6. Elston J T 1992 Heysham 2 project case history. *International Journal of Project Management* **10**(3), August 179–84
7. Barnes M 1986 *The CESMM2 Handbook* Thomas Telford
8. Ridout G 1982 Target cost takes the risk out of contracting. *Contract Journal* 14 October: 16–18
9. Stukhart G 1984 Contractual incentives. *ASCE* **110,** March: 34–42

Further reading

Graham P 1987 Canary Wharf and the bashed B of Q. Letters to the Editor, *Building* 20 November: 28

Kinder, D C 1986 *An Appraisal of the Target Cost Contract* MSc Construction Management project report, Loughborough University of Technology

7

Tendering procedures and bid evaluation

Introduction

In normal circumstances the purpose of an invitation to tender is to obtain from the contractor a firm offer capable of acceptance and hence conversion into a binding contract. The purpose of assessment is to carry out a comparison of the offers received with the intention of entering into a contract with the successful tenderer.

The client can control the selection of contractors at the pre-tender stage, i.e. before the issue of the tender documents and secondly before the award. The principal choice at the pre-qualification stage is whether to adopt a full pre-qualification procedure specific to each contract or whether to develop standing lists of suitably qualified contractors for various sizes of contracts and types of work.

Most public sector authorities and many large private clients have their own standing orders and procedures which in the main are based on the industry's standard procedures. It is particularly important that standard procedures are used in order that the parties know where they stand in the event of any problems occurring.

Qualified bids and errors in tenders are not uncommon on major projects. The contract administrator should therefore have a standard policy for dealing with such items and should not be drawn into a 'horse-deal' with the lowest tenderer before the project has commenced. Such a deal could result in one party feeling aggrieved from the start which could have adverse repercussions throughout the project.

In this chapter we examine the standard procedures in the building and civil engineering sectors in the UK and overseas. In both sectors contracts are traditionally awarded to the responsive bidder with the lowest price. However, enlightened contract administrators are aware that this system will not necessarily ensure that the client obtains the lowest final account, or the best value for money, or is necessarily the lowest bid when the time-value of money is considered. The optimum bid is the lowest priced bid which has undergone a process of assessment to identify and, where necessary to price, the consequences inherent in the submission.

European Union tenders

The single European market has, in theory, created a potential domestic construction market of £420bn. a year. Access to this single market operates through a system of directives which are intended to allow free competition to public contracts throughout the European Union. The directives require that a contract notice is published, free of charge, in the *Official Journal of the European Communities*.

There are two main directives relating to construction. The Public Works Directive 89/440/EEC of 18 July 1989 which amended Directive 71/305/EEC of 26 July 1971. This directive requires a contract notice to be given for all building and civil engineering projects by public bodies over a value of ECU 5m. (£3.75m.).

Likewise the Utilities Directive 90/531/EEC which covers the procurement of supplies and works by public and private purchasers in the energy, water, transport and telecommunications sectors requires a contract notice to be given for all work over a value of ECU 5m. (£3.75m.).

Information on public building and civil engineering projects throughout the European Union are categorised in the daily *Official Journal* under the headings:

- pre-information procedures;
- open procedures – all interested contractors may submit tenders;
- restricted procedures – only those contractors invited by the contracting authority may submit tenders;
- accelerated restricted procedures;
- notice of public work concessions and contract awards.

The Public Works Directive allows for negotiated tenders only in certain circumstances, e.g. unforeseen events requiring emergency action, or additional works to an existing contract. In contrast the Utilities Directive contains no such restriction on the use of negotiated tenders.

Building sector – single-stage selective tendering

The most appropriate method for obtaining tenders in the building sector is by means of single-stage selective tendering. The standard procedures are identified in the NJCC *Code of Procedure for Single Stage Selective Tendering*[1] which is designed for use with JCT 80, the Intermediate Form or the Minor Works Agreement. Provisions of the code are qualified by the supplementary tendering procedures specified in the European Council directives.

Once it has been decided that a selective tendering system is to be used tenderers should be drawn up either from the employer's approved list or from an *ad hoc* list of suitable contractors. The code recommends that the list of tenderers is limited to a maximum of six, with one or two further names to replace any withdrawals.

The criteria that should be considered when selecting the short list are:

1. The firm's financial standing.

2. Whether the firm has recent experience of achieving the required productivity rates over a similar period.
3. The firm's general experience and reputation in the specific field.
4. Whether the firm's management structure is adequate for the type of contract envisaged.
5. Whether the firm will have adequate capacity at the relevant time.

The employer's approved list should be periodically reviewed and those firms whose performance has been unsatisfactory should be removed. Contractors should be invited to tender on a rotating basis, e.g. contract A – firms 1 to 6, contract B – firms 7 to 12.

Each contracting firm should be sent a preliminary invitation to tender, 4–6 weeks before the receipt of the tender, in order to decide whether or not to tender.

The Code of Procedure strongly recommends that standard forms of contract are used in an unamended form. A minimum period of 4 weeks should be allowed for the preparation of tenders, with major projects requiring a longer period.

All tenders should be based on identical tender documents and tenderers should not attempt to vary that basis by qualifying their tenders. Any tenderer who submits a qualified tender should be given the opportunity to withdraw the qualifications without any amendment to the tender; if not withdrawn the whole tender should be rejected if it is considered that such qualifications afford the tenderer an unfair advantage over the other tenderers.

The code further recommends that after the tenders are opened, the lowest tenderer should be asked to submit the priced bill of quantities (B of Q) as soon as possible and no later than 4 days. In practice priced B of Q are often required to be submitted by all the tendering contractors.

All but the lowest three tenderers should be informed immediately that they have not been successful. The second and third lowest tenderers should be notified that they were not the lowest but may be approached again if it is decided to consider their offers. Once the contract has been let every tenderer should be notified of the tender prices.

Section 6 of the Code of Procedure identifies two alternatives in the event that the quantity surveyor detects any errors in computation of the lowest tender. The preliminary invitation to tender should state which of the alternatives should apply.

Alternative 1 requires that the tenderer be given such details of errors and afforded an opportunity of confirming or withdrawing the offer. If the tenderer withdraws the second tender should be examined.

If the contractor confirms the offer an endorsement should be added to the priced bills indicating that all rates (excluding preliminary items, contingencies, prime cost and provisional sums) are to be considered as reduced or increased in the same proportion as the corrected total of priced items exceeds or falls short of such items.

The corrected B of Q will be used for the financial administration of the contract and a percentage reduction or addition should be applied to all those sums established using B of Q rates both in the interim valuations and variations.

Alternative 2 requires that the tenderer should be given the opportunity of confirming the offer or of amending it to correct genuine errors. If the decision is to amend the offer the revised offer may no longer be the lowest. If it is decided not to amend the offer an endorsement will be required as in alternative 1.

The code further recommends that the contractor be allowed sufficient time for pre-planning the project (not exceeding 2 months) prior to commencement on site.

Building sector – two-stage selective tendering

In some circumstances, particularly on large or complex projects, it may be appropriate to appoint the contractor early and use the two-stage selective tendering procedure, e.g.

1. When the contractor can make a technical contribution and become involved in the planning during the pre-construction process, or
2. If the employer wishes to start on site before all production information is available due to a tight programme.

The NJCC *Code of Procedure for Two Stage Selective Tendering*[2] assumes that the employer's professional team retains responsibility for the design throughout. However, the code acknowledges that in practice much of the specialist design work is carried out by specialist subcontractors. The code recommends therefore that there should be a separate direct agreement with the employer for such design work and the subcontractors should be appointed by the nomination procedure as laid down in the standard building forms.

The first stage of the system involves the selection of a contractor by means of a competitive tender based on pricing documents related to preliminary design information and which provide a level of pricing for subsequent negotiations.

The nature of the first-stage documents must depend on the circumstances of the individual project. If the design is sufficiently advanced bills of approximate quantities may be used. Failing this some other method which requires the tenderers to provide an analysis of their 'all-in' labour rates, percentage additions for profit and overheads, details of subletting and plusages for profit and attendances and a detailed build-up of preliminaries.

The second stage of the process involves the finalisation of the design by the employer's design team in conjunction with the contractor and the preparation of a B of Q priced on the basis of the first-stage tender. Upon agreement an acceptable sum is established for incorporation into a standard form of building contract, e.g. JCT 80 with Quantities.

Where the employer fails for any reason to reach agreement with the selected contractor it will be necessary to recommence second-stage procedures with the next tenderer or reinvite tenders. However, this course of action could seriously delay the programme and adversely affect the cost.

On 3 June 1988 *Building* magazine featured an excellent article on the construction of the £60m. Royal Mint Court speculative office development in London.[3] The 55 000 m² project was completed under a two-stage tendering system barely 2 years after the developers acquired the site. Six contractors were given 3 weeks to submit first-stage tenders comprising:

- a mark-up on net cost items;
- a construction programme;
- a statement of construction method;
- the track record of the firm and staff involved.

Within one month Laing Construction had signed a £500 000 preliminary contract of enabling works comprising removing overburden and demolition. The rates were established based on the same mark-up of prime cost that the contractor had submitted for the main works.

Whilst the contractor was carrying out the enabling works the quantity surveyor was drawing up the B of Q. Three months after the preliminary contract was signed Laing submitted its second-stage tender and after negotiation was duly appointed as main contractor on the basis of a fixed price lump sum and tender period.

The lump sum agreed was for the basic trades of groundworks, concrete, brickwork and decorating. At later dates specialist subcontractors were asked to tender on separate packages of work and were then adopted by Laing as domestic subcontractors.

The Royal Mint Court project was based on the JCT 80 form of contract with certain amendments relating to co-operation and input on buildability. The aim was to introduce constructive thought rather than entrenched positions; obviously a system that relies on a certain amount of confidence and respect between the parties.

Civil engineering sector – procedures within the UK

General

A well-established reference for measure and value civil engineering works is the *Guidance on the Preparation, Submission and Consideration of Tenders for Civil Engineering Contracts*.[4] However, many of the procedures will be applicable to other forms of admeasurement engineering contracts such as pipework and cabling.

Public accountability may require acceptance of the lowest bid but the reasons for not accepting such a bid have been identified as:

- no guarantee that the lowest bid will be the lowest final account;
- lowest bid may be suicidally low or misconceived;
- lowest bid might not be most realistic.

The lowest bid is therefore the lowest priced evaluated bid which has undergone a process of assessment to identify and, where necessary, to price the consequences inherent in the submission.

Engineering contracts, by their very nature, provide opportunities for the submission of *qualified tenders* or *alternative designs* for part of the works.

The standard procedures must provide for these eventualities in a manner which is equitable to all the parties concerned.

Open tendering is not recommended; selective tendering based on approved lists or pre-qualification is recommended. Contractors invited to pre-qualify should be asked to submit details under the general headings: contractor's financial standing, technical and organisational ability, general experience and performance record.

The number of contractors invited to tender is recommended as not less than four and no more than eight; as a general rule the larger the project the fewer the number of tenders invited. The aim should be to invite only those contractors who have the necessary technical and financial resources to complete the contract satisfactorily.

Preliminary enquiry

Preliminary enquiries as to a firm's desire to tender should be sent to the selected contractors in order that the required number of tenders is returned and any contractor declining to tender can be substituted. Such enquiries should contain comprehensive details including: the nature of the work, the major quantities, the commencement date, any special features and the order of the cost. It should be made clear to contractors that their refusal to tender will not prejudice their opportunity for tendering on future occasions.

Tender documents

Following acceptance of the preliminary enquiry contractors should be sent the tender documents together with the 'Instructions to tenderers'. These instructions outline the contractor's obligations, explain the format of the information to be submitted to the client, describe the terms of employment, working practices, procedures and methods of payments. The instructions should also attempt to minimise the submissions of qualified bids and provide details of procedures which should be adhered to in the event of problems occurring.

All information on the site and the ground conditions should be included in the tender or made available to contractors tendering. It is recommended in the guide that 4 weeks should be allowed for tendering or longer on a major or complex project; in practice the period allowed for the submission of bids varies between 6 and 18 weeks.

If any queries of any significance are received during the tender period the engineer's response should be circulated to all the tenderers.

Tender period

A *pre-tender meeting* and/or *site visit* of all tenderers may be utilised in order to clarify all anomalies and attempt to eliminate qualified tenders. However, care must be taken not to destroy the element of competitive tendering or disclose one contractor's unique solutions to the others.

Following the meeting all the information given should be confirmed to all tenderers.

Tenders received after the due time and date for submission should be rejected. However, a late tender may be considered if a cable, telex or fax stating the tender sum is received on time and there is clear evidence that the complete tender documents have been dispatched at a reasonable time.

The 'Instructions to tenderers' should state whether alternative designs will be considered. Such alternative designs, should take into account any of the engineer's special design criteria and requirements and should be considered only provided a 'clean' tender based strictly on the tender documents is also submitted. The alternative tender should be accompanied by supporting information such as drawings, calculations and priced B of Q in order that it can be fully assessed.

As well as including design alternatives tenderers may include alternatives for the construction programme or methods and sequence of work. If the proposal involves shortening the contract period the engineer must consider whether he will be able to produce the information to the contractor at the required time. Furthermore it is vital that any examination of the alternative proposals does not delay the commencement of the works.

In practice an alternative bid would normally be considered only if it was technically, contractually and financially acceptable. Unless there were substantial cost or time savings offered to the client it is likely that an alternative bid would be rejected at the initial evaluation stage.

Any qualifications made by a contractor in his tender will undermine the principles of parity of tendering. The engineer can minimise the submission of qualified tenders by careful pre-planning and by holding a pre-tender meeting.

However, in some circumstances it may be in the best interest of the employer for the engineer to consider and evaluate a qualified tender. It is therefore essential that the 'Instructions to tenderers' clearly states the circumstances which will lead to a tender being rejected, or otherwise how they will be dealt with. All qualifications should be considered in detail and their true financial implications established.

Tender adjudication

It is good practice to describe in the 'Instructions to tenderers' the procedures that will be adopted to deal with errors. The first check to be made is the arithmetical accuracy of the bid. The contractor's rates are the only firm element in the bid as the works are subject to admeasurement. Any arithmetical errors in multiplication, totalling, transferring to collections, etc. should be corrected and it will be the corrected totals which form the basis of the tender adjudication.

The engineer should also identify any discrepancies between each bid and with the client's estimated rates. He should establish whether the bid is front-end loaded or contains significantly high rates for items whose quantities are likely to increase, e.g. excavation of rock.

The engineer should further check for inconsistencies between rates with the same description in different sections of the B of Q. Any inconsistencies should be clarified before the acceptance of the tender; these observations can be particularly relevant when negotiating the price of post-contract changes.

The 'Guide' recommends that it is not regarded as good practice to negotiate for a change in a tendered rate on the grounds that it is in error or inappropriate. However, in practice the engineer might request the contractor to reallocate those excessive rates without altering the overall bid price.

Merna and Smith in their 2-year SERC research study into the different bid evaluation procedures used in the UK public sector civil engineering work[5] found that the assessment of bids is based largely on the B of Q (Table 7.1). However, they discovered that many clients who were concerned with operations as well as construction rely increasingly on the submission of non-contractual information to try to identify likely performance. They further found that some clients also compared the estimated cash flow projection with those of the tender bids, occasionally using a discounted cash flow method.

Following the submission of all the tenders a *post-tender meeting* will be held with lowest tenderer and possibly the second or third lowest. Any relevant points which have been identified in the tender scrutiny should be discussed, approved by both parties and recorded in writing for incorporation into the contract documents.

At the conclusion of the adjudication process the engineer should submit an assessment report to the employer setting out the procedures undertaken to check and correct the tenders and giving his recommendation to the employer.

Acceptance procedure

Once the contract has been let all the tenderers should be supplied with a list of the names of all the contractors tendering in alphabetic order and a separate list of the tender totals submitted in ascending order of magnitude. The time taken to evaluate the submitted bids is not usually specified but most take between 10 and 12 weeks.

Merna and Smith[5] considered that a more economic system could be implemented without adversely affecting the client yet at the same time saving in valuable engineering resources. This wasted resource is caused by each contractor's estimator being required to price the detailed B of Q after building up their prices using the operational tendering approach; this is particularly significant on major projects.

The conclusions of their research recommended a procedure that would require the client receiving a tender price total from six contractors, a detailed B of Q only from the lowest three bids and a pre-award meeting with the representatives from the lowest contractor only.

Table 7.1 Pre-qualification and evaluation methods used by UK public clients on civil engineering works

| Client | Prequalification | | | Number of tenderers | Tender period | Bids | | Pre-award meeting | Evaluation | |
	Origin	List of tenderers	Other			B of Q	Other		Other	Price
Department of Transport	Advertisement Approved list	30–40	—	4–8	Up to 12 weeks	All	—	Yes	Time	Lowest conforming
Property Services Agency	Approved list	Varies	Competition Location Previous work	Fixed to value 4–10	Up to 8 weeks	—	Tender sum	Sometimes	—	Lowest conforming
British Waterways Board	Approved list	—	Location Previous work Distribution	6–8, varies	—	All	Method statement Programme	Yes	Time	Lowest conforming with time
British Nuclear Fuels	Records	Varies	Interviews	Varies with price	4–12 weeks	All	Method statement Programme Production plant scheme	Yes	Security Quality	Lowest conforming
British Rail	List	Large list	Previous work	4–8, varies	3–4 weeks	All	Programme Safety	—	Time	Lowest conforming with time
British Gas	Approved list	—	Previous work	4–8, varies	—	All	Method statement Programme	Sometimes	—	Lowest conforming
British Coal	Records	—	Location Previous work	4–8, varies	4–8 weeks	All	—	Yes	Time	Lowest conforming
British Airports Authority	Records	Varies	Previous work	6–8, according to value	Up to 6 weeks	All	Method statement Programme	—	Time Safety	Lowest conforming
Yorkshire Water Authority	Records	40–50	Competition	4–10, varies	Up to 10 weeks	All	—	Sometimes	—	Lowest conforming
Central Electricity Generating Board	Records	Varies	Project size	4–8, varies	12–16 weeks	All	Method statement Programme	Sometimes	Management Safety Quality	Lowest conforming
Gwent County Council	Advertisement	Varies	Local competition	6	4–13 weeks	All	—	—	—	Lowest conforming

Source: Merna and Smith.[5] Reproduced by permission of the Institution of Civil Engineers.

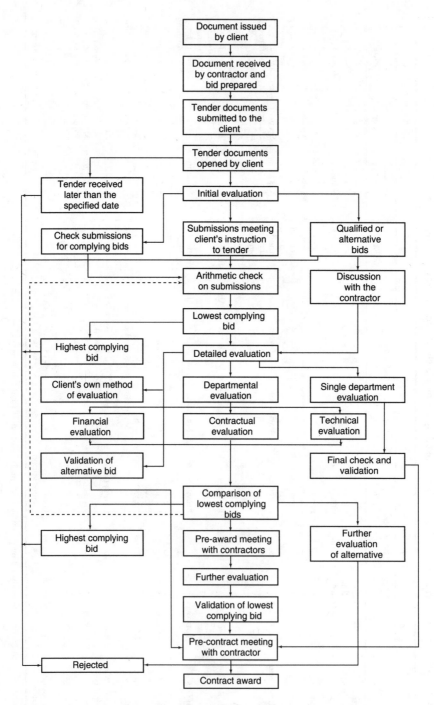

Figure 7.1   Typical evaluation procedures. *Source:* Merna and Smith.[5] Reproduced by permission of the Institution of Civil Engineers

Civil engineering sector – international procedures

Many of the guidelines identified in the FIDIC document *Procedure for Obtaining and Evaluating Tenders for Civil Engineering Contracts*[6] are similar to those already described above under UK civil engineering procedures. Interestingly, the authors of the FIDIC document consider that not only are the procedures relevant to projects undertaken under the FIDIC Conditions of Contract but they can also be related to any acceptable contract form; hence our brief examination of the document.

Not surprisingly, there is a large section of the FIDIC document devoted to the pre-qualification procedures which are considered essential for large and/or complex projects.

The 'Evaluation of tenders' section in the FIDIC document requires that appropriate key factors and method of evaluation should be established in advance so that subsequent evaluation and comparison of tenders leads to an objective judgement.

The FIDIC guide considers that the evaluation of tenders can generally be considered to have three components which are brought together for total analysis and judgement. These components are identified as:

1. Technical evaluation:
   (a) conformity with specifications and drawings
   (b) comparison of any proposed alternatives (if allowable) with the requirements of the enquiry
   (c) methods of construction and temporary works
   (d) programme
   (e) subcontract work
   (f) construction plant and equipment, etc.
2. Financial evaluation:
   (a) capital cost
   (b) discounted cash flow and net present value
   (c) programme of payments
   (d) financing arrangements
   (e) currencies
   (f) bonds, guarantees
   (g) interest rates
   (h) down payments/retentions
   (i) daywork rates
   (j) contract price adjustment proposal, etc.
3. General contractual and administrative evaluation:
   (a) conformity with instructions to tenderers and enquiry
   (b) completeness of tenders
   (c) validity of tenders
   (d) exclusions and qualifications (stated or implied)
   (e) insurance
   (f) administrative expertise
   (g) shipping, customs, transport
   (h) working hours
   (i) labour build-up, run-down and source, etc.

The tender with the lowest sum should be judged the most advantageous offer financially provided that the technical, contractual and administrative aspects are satisfactory.

The FIDIC guide suggests that in the final stages of the appraisal a risk analysis should be made of possible consequences to the employer of inadequate quality and lack of performance by the contractor if these situations have not been adequately covered in the contract documents. This risk analysis is particularly important if the contract is large and complex and should be considered under the following broad headings:

- failure to meet programme completion dates
- failure of temporary or permanent works or elements of the work
- lack of performance of mechanical and electrical plant and services, if any
- effects of lack of collaboration with others closely allied to success of the project.

The section 'Evaluation of tenders' in the FIDIC guide concludes with the statement: 'For any project, the lowest tender in the financial analysis may not necessarily be the best tender for the employer to accept and a balanced view must therefore be taken of all the factors embraced by the evaluation.'

Developing a method for tender evaluation

On large contracts, or in the case when two or possibly three tenders are particularly close in price, it is preferable to try to quantify which is likely to give the client best value for money.

Many clients/project managers claim reasonable success by using a points system. The strength of such an approach is that it focuses attention on all the components of the bid as well as the whole, it attempts to quantify parameters which are qualitative and it provides a sound basis on which to justify the decision. However, weaknesses in the approach could include the possibility of conflicting views when establishing the key factors, the fact that many factors cannot be evaluated financially and the scoring system may provide a harsh strait-jacket.

One possible approach is for the engineer or contract administrator to develop two lists of criteria; one for general factors and the company's past performance and the other for specific factors on the current bid. The list of items should be reviewed for each project and each item should be weighted depending on its significance to the project; this concept is based on the matrix scheme described by Glen Peters in his book *Project Management and Construction Control*.[7]

General factors on company/ past performance		Specific factors on current bid	
1. Financial standing	20	1. Qualifications and expertise of key personnel proposed	20
2. Experience and past performance of similar work	20	2. Own key labour	10
3. Quality management systems	15	3. Major items of own construction plant	10
4. Reputation for achieving programme completion	15	4. Programme proposals	10
5. Reputation for litigation/claims	20	5. Method statement and temporary works	20
6. Safety record and policy	10	6. Contractor's pricing strategy	20
		7. Adequate capacity	10
	100		100

The three lowest bids on the project after adjusting for all tender qualifications are found to be: contractor A £55 460 000, contractor B £55 485 000, contractor C £55 490 000.

However, an analysis of each contractor's general/past performance and specific performance shows the following:

	General							Specific							Total
	1	2	3	4	5	6		1	2	3	4	5	6	7	
Contractor A	10	10	5	10	5	5	45	10	5	5	10	10	10	0	50	95
Contractor B	15	15	15	15	10	10	80	15	10	10	10	20	15	10	90	170
Contractor C	15	10	15	15	15	10	80	15	10	5	5	15	10	10	70	150

	Potential high risk	Acceptable	Recommended
	100	150	200

Contractor A************
Contractor B***
Contractor C**********************************

If the price alone is the determining factor then contractor A would be selected; however, contractor B would seem better choice.

Hawwash[8] considered that a possible refinement on the bid evaluation by points system might include:

1. Using points to establish a minimum acceptable level. Decision based

on price alone for all bids above this level (this may be inappropriate where performance/time dominates cost).
2. Assessing in advance an equivalent price for a certain increment of qualitative points (i.e. a differential between bids of x points is equated to £y).
3. Weighting price and qualitative points, e.g. final decision based on 60 per cent price + 40 per cent points. A method of relating prices to a scale of 0–100 is then needed.

An alternative approach is for the engineer to request two sealed envelopes from each contractor; one containing the technical submission, i.e. method statement, resources, programme and the other containing the financial bid. The engineer initially opens the technical submissions and recommends the three most appropriate, then the tenders from the same three contractors are examined and the awards made to the lowest tenderer. This method avoids giving the project automatically to the cheapest contractor and yet can be justified as a sound approach.

Case study

An examination of the HKMTRC pre-contract appraisal system indicated the following procedure. After a detailed examination of the tenders the reports listed below should be submitted in person to the employer's executive by the key members of the PM team.

Finance

After considering the terms of any financial offer from the contractor, e.g. deferred payment terms, examining for front-end loading, making an arithmetical check, calculating the net present value of the bid and the contract cash flow, the report should include:

- preliminary financial assessment
- summary of the financial offers
- summary of alternative tenders
- identification of the more favourable bids in finance terms

Contracts/legal department

After examining the tenderers' covering letters and all qualifying statements the report should indicate the cost and programme implications of tenderers' qualifications and should advise other group members where the actions lie.

Programming

The report should indicate any deviations from the schedules of milestones/critical dates.

Consultants/civil engineering department

This report should identify any alternative design offers and recommend any which warrant further consideration and should advise of any planning implications. It should examine the method statement from tenderers and highlight any anomalies and areas of non-acceptance. A detailed cost centre comparison should be prepared to enable cost centres and activity bills to be compared with the engineer's estimate.

Utilities and planning

The report should evaluate the implications of the tenderers' submissions on traffic, land access and utility diversions.

Following a detailed study group members should prepare questions to be put to tenderers in order to clarify any points in the tenders and remove all qualifications. These questions are then edited and sent to the contractors.

Following receipt of the answers negotiation sessions are held with the tenderers and all outstanding matters are clarified. A 'letter of clarification' or 'wrap up' letter is then prepared and incorporated into the contract.

Conclusion

We have now examined in detail the tendering and bid evaluation procedures in the building and civil engineering sectors. We have found that many major employers are aware of the weaknesses inherent in awarding the contract to the lowest bidder and we have identified the issues to be considered in the appraisal procedure.

In the early 1980s Hardy, Norman and Perry at UMIST carried out research for the World Bank in order to develop a practicable procedure for the evaluation of bids for construction using discounted cash flow techniques. A paper was published[9] which identified that there may be advantages to the client if:

- bidders were invited to specify their preferred duration of construction;
- contractors were invited to specify the magnitude and timing of the mobilisation;
- expected patterns of payment for measured work were compared and the effect of price escalation on different bids was considered.

The UMIST team further stated that bidders should be required to submit a schedule of payments which they expected to fall due to them under the contract. Present values for bid comparison would be calculated from the schedules in a manner described in the bidding documents. The bid payment schedule was to be used only for bid evaluation and had no contractual significance once the contract was awarded; this was mainly due to possible administrative problems in the handling of extra works and extensions of time. The UMIST teams considered that further study was necessary into cases where prescribed payment schedules were made contractual.

In 1988 the author described such a scheme, where the prescribed payment system was made contractual, in his article on an alternative payment system for major fast-track construction projects.[10] The system described, which utilises conventional B of Q, has been used by the HKMTRC and has been found to be a success. However like all successful systems, and indeed projects, its success very much depends on the spirit of co-operation between the parties.

It can be seen that current practice in contractor selection is fragmented and can be subjective. At the University of Wolverhampton a research project is currently (1993–5) being undertaken developing an alternative quantitative technique for the selection of contractors. This technique when developed should furnish the PM team with a decisional aid in the form of probability scores at three critical stages in the selection process: pre-qualification, tender evaluation and final selection. Much interest in research project is being shown by practitioners in industry. Reference 11 describes the research in detail.

Tutorial questions

1. (a) Single-stage selective tendering usually produces tenders which are more expensive than those gained in open competitive tendering. Why, therefore, is competitive tendering not used more frequently?
    (b) There are tendering methods which bridge the gap between single-stage selective tendering and open competitive tendering. Explain these other methods and indicate the circumstances where they may be appropriate.

    (RICS Final Quantity Surveying (1987 syllabus), specimen paper)

2. Tenders received on a further phase of a leisure park development were as follows:

Tender	Price
1	£23 110 000
2	£26 020 000
3	£27 220 000
4	£28 350 000
5	£35 320 000
6	£40 285 000

    The firm price tenders were based upon B of Q from six different contractors competing on a selective basis. The budget was £30m.

    Discuss what factors can lead to such variation in tender prices. What would your recommendations be to your client? (RICS Final, Graduate Entry Scheme (Final) Examination 1990 (1987 Syllabus) Paper 3)

3. Six contractors are expected to submit tenders for leisure park self-catering units on Friday 16th (assume today is the 8th). The architect has been contacted by an additional contractor, asking if his firm could also submit a bona fide tender. The contractor has indicated that a 2-week extension may be required, or alternatively the contractor could prepare a tender for negotiation in competition with the submitted tenders. The architect feels that this would almost certainly make for a lower tender sum.

(a) Explain the ethos of competitive selective tendering (9 marks).
(b) Discuss the potential value of negotiation in conjunction with competitive tendering (8 marks).
(c) Describe an appropriate course of action for the architect. (8 marks).

(RICS Final, Graduate Entry Scheme (Final) Examinations 1990 (1987 syllabus) Paper 4)

4. You are the engineer for a project under the ICE Conditions of Contract measured in accordance with CESMM2 for which tenders have been received from six selected tenderers.

    Describe the checking procedures you would carry out before recommending a tender to the employer for acceptance, indicating in particular how you would deal with errors of rates and what steps you could and should take to ensure that the aggregate interim payments made to the contractor in accordance with the provisions of the contract at any stage during the construction of the works would be a fair valuation of the work done. (ICE Examination in Civil Engineering Law and Contract Procedure June 1987, Paper 3)

5. Describe the various enquiries normally made by promoters and engineers to arrive at a select list of contractors who will be invited to tender for a project.

    List the information which you consider should be included in instructions for tenderers for a bridge project.

    Eight tenders based on a remeasurement B of Q are received for the project. One of these, in addition to a conforming offer, contains a lump sum alternative which is some 10 per cent lower than the lowest offer. Another lump sum offer which is not accompanied by a conforming offer is 15 per cent lower than the lowest conforming offer. As the engineer, write a report to the promoter recommending a tender with your reasons. (ICE Examination in Civil Engineering Law and Contract Procedure June 1988, Paper 3)

6. Prepare a draft pre-qualification questionnaire to be sent from the client to contractors wishing to tender for a major £50m. design and build urban motorway project, clearly identifying the main issues involved

References

1. National Joint Council for Building 1989 *Code of Procedure for Single Stage Selective Tendering* NJCC Publications, April
2. National Joint Council for Building 1983 *Code of Procedure for Two Stage Selective Tendering* NJCC Publications, January
3. Spring M 1988 A tender two step. *Building* 3 June: 40–4
4. Institution of Civil Engineers, Association of Consulting Engineers and Federation of Civil Engineering Contractors 1983 *Guidance on the Preparation, Submission and Consideration of Tenders for Civil Engineering Contracts Recommended for Use in the United Kingdom*
5. Merna A, Smith N J 1990 Bid evaluation for UK public sector construction contracts. *Proc. Instn Civ. Engrs*, Part 1, **88** (February): 91–105

6. FIDIC 1982 *Tendering Procedure: Procedure for Obtaining and Evaluating Tenders for Civil Engineering Contracts*
7. Peters G 1981 *Project Management and Construction Control* Construction Press
8. Hawwash K 1992 Selection of contractors and tender analysis, Lecture notes, Advanced Project Management course, UMIST, March
9. Hardy S C, Norman A, Perry J G 1981 Evaluation of bids for construction contracts using discounted cash flow techniques. *Proc. Instn Civ. Engrs* Part 1, **70** (February): 91–111
10. Potts K F 1988 An alternative payment system for major 'fast track' construction projects. *Construction Management and Economics* **6:** 25–33
11. Holt G D *et al.* 1994 Evaluating performance potential in the selection of contractors *Engineering, Construction and Architectural Management*: **1/1** 29–50.

Further reading

Fish G F 1985 Tendering in a competitive market. *Chartered Quantity Surveyor* August: 23

Knight H 1979 Selecting the civil engineering contractor. *Chartered Quantity Surveyor* July: 204–6

Renwick T, Burton T 1994 European directives. *Building* 4 February: 40–1

Section D

Management of the tendering stage

8

Contractor's estimating and tendering

Introduction

The submission of successful tenders is obviously crucial to the very existence of contractors. Yet a fundamental truth of competitive tendering, particularly on major works, is that the lowest tenderer is often the one who has most seriously underestimated the risks which obviously could have drastic consequences particularly in times of recession when margins are slim to say the least.

Despite this truth, however, there can be some scope for the innovative contractor when tendering. It is not unknown, on major civil engineering projects, for the award to be made to the contractor who has devised a more economic design than the one proposed by the engineer for part or even the whole of the works. Furthermore, contractors have been awarded contracts even though their tenders were not the lowest; this may be due to a highly original method statement and design solution for the temporary works, or possibly the offer of a deferred payment scheme.

However, before any estimates can be submitted the first step for contractors is to get on to tender lists. This could be done on an *ad hoc* basis or preferably in accordance with some sort of longer-term strategic marketing plan. The plan, which should be based on an analysis of the past and a consideration of the future trends within the market, should be re-examined on an annual basis and modified accordingly.

Stage 1 – Decision to tender

The first stage in the tendering process is the decision to tender. As soon as the tender documents are received the estimator should quickly skim through the documents in order to establish:

1. The amount and type of work involved and whether the company has any competitive advantage.
2. The approximate value together with a review of the major resources required for the project, particularly construction equipment, staff, key subcontractors and suppliers.

3. The programme requirements, i.e. completion, sectional completion and critical milestone dates.
4. The form of contract, specification, method of measurement and if there are any amendments made to the standard documents, e.g. deletion of grounds for extension of time.
5. The time and resources required for preparation of the tender.
6. Whether any contractor's design is required, and whether the main contractor is required to accept liability for subcontractors' design.
7. Possible alternative methods of construction and temporary works.
8. Whether the risks are acceptable and whether the tender is of particular interest.
9. Contract requirements for performance bonds, warranties and parent company guarantees.
10. Funding requirements for project (based on cash flow forecast – comparing cash in/cash out).

Upon completion of the review the estimator should complete a pre-tender data sheet, grade the tender based on the interest to the company and recommend whether or not to tender. If a contractor decides not to tender the documents should be returned to the engineer; however in practice this rarely occurs.

The technical process of predicting the net cost of the works is carried out by a team comprising the estimator, planning engineer, materials estimator, estimating technician, together with possible contributions from temporary works designers and an experienced construction manager if the work is of a specialist nature. At the end of this process the team will produce the *cost estimate*.

Stage 2 – Determining the basis of the tender

During this stage the estimator, prior to the preparation of the cost estimate, will disseminate and assemble the key information and generally become familiar with the documents. Unlike pricing a bill of quantities (B of Q) in the building sector, a civil's B of Q can be priced only when read in conjunction with the engineer's drawings and the specification. Projects carried out for the water authorities, British Rail and highway works for the Department of Transport are normally in accordance with standard specifications.

Enquiries will be sent to subcontractors and major materials suppliers, the latter often based on the quantities calculated by the contractor's quantity surveyor from the drawings. The contractor should also check that the major quantities in the B of Q are correct; if any are found to be incorrect this factor will be considered later at the commercial appreciation stage.

However, the most important part of this stage is for the team to determine the *construction method* and sequence upon which the tender is based, together with an *outline programme of the works*; two items that are inseparable.

The construction method will often be dependent on the design of the temporary works necessary to enable the permanent works to be constructed. Temporary works are normally designed in-house by the contractor but, in the case of scaffolding and falsework design, may be supplied by specialist contractors.

Temporary works may have considerable time and cost implications, and can include cofferdams and temporary access bridges on riverworks, temporary piling and jetties on marine projects, overhead gantries on elevated motorways, dewatering systems and grout curtains on deep basements, etc.

The contractor may further be required to design part of the works to meet a performance specification, for example concrete specified to a strength or piling to a load-carrying capacity; this design would often be undertaken by specialist suppliers or subcontractors.

A further involvement of the contractor's design department may be in identifying more economic alternative solutions for sections of the permanent works. This is often done in the hope of sharing the saving involved, which could be considerable.

The programme could be in the form of a bar chart or in the case of major works based on a network showing the critical path produced utilising computer techniques. The programme will be used by the successful contractor as a control document for monitoring progress and calculating the effects of any delays and disruption to the flow of the works.

The programme is particularly important as 15–40 per cent of the cost of civil engineering works is time related and many items such as site overheads are computed directly from it. Furthermore, as most contractors will be bidding using the same quotations from subcontractors, plant hirers and materials suppliers, obtaining a saving in time is one of the few ways in which the contractor can show a substantial saving to the project cost.

During this stage the estimator will further need to identify any inherent restrictions (e.g. delivery of materials by rail or water) and any items on long deliveries (e.g. specialist equipment). He will further need to consider alternative methods of construction, sequence of construction and the level and utilisation of resources.

In accordance with the requirements of the clause 11(2) of the ICE 6th the contractor is deemed to have inspected and examined the site and its surroundings. The contractor may also visit the engineer's office and the local authority in order to examine core samples, existing services, traffic requirements and any other information available.

Following the site visit a comprehensive standard pro forma checklist will normally be prepared listing such items as: access to site, site security, provision of services, soil information and groundwater, nature of excavation fill and disposal, nearest tipping facilities, availability of labour, construction equipment and materials, site organisation and layout, land purchase for borrow pits, etc.

The method statement is a key document in the preparation of the tender and should consider the site visit report, the geotechnical report, the sequence and methods for the main operations of work, subcontracted work, bulk quantities, schedules of labour and construction equipment and any temporary works required.

If the method statement is required to be given to the client, either with the tender or in accordance with clause 14 of the ICE 6th, many contractors – if they have a choice – choose to submit a statement covering only general principles, i.e. not too detailed.

Typical contractor's method statement (for submission to client)

(based on actual example)

Baker Street Station Footbridge

This method statement is to be read in conjunction with our Programme No. 3000/T/PLAN/90094.

Suitable hoists, site offices and hoardings will be erected at commencement at positions/areas indicated on Drg. No. 797/90.

Excavation work will be carried out using suitable mechanical plant with due regard for safety and stability. Trench boxes or braced heavy-duty trench sheets would be used to support excavation sides. Every attempt will be made to hand back the platforms with required minimum width, but due to the layout and configuration of the bases this may not be possible every time.

Arrangements for distribution of concrete for the bases will be made from either the Pallisades or Baker Street. On completion of the concreting at any location, the permanent columns and temporary works required to accept the launch of the bridge will be erected. Sections of the staircases requiring erection underneath the bridge, will be pre-assembled and positioned prior to launch.

The bridge steelwork will be assembled in sections and part clad at Loam House and and moved on low loaders to Baker Street, for completion of the cladding. Before this the street kerbs, traffic lights, streetlights, etc. will have to be altered in conjunction with the City Council to facilitate the transportation during road closures.

The bridge fully clad will be launched in two sections from Baker Street and one section craned from Low Street to detailed method statement to be prepared by our consulting engineers. The principle has already been discussed with British Rail in substantial detail and is outlined in our attached Schedule of Proposed Occupations.

The remaining sections of staircases, part assembled steelwork and kites will then be craned into position and fixed.

The cladding to the staircases and kites will be carried out using boom hoists.

Work on the construction of the concourse will commence on completion of the launch.

The required possessions in Phase B of our programme and work to be undertaken in such possession is attached herewith as Schedule of Proposed Occupations.

Stage 3 – Preparation of the cost estimate

During this stage the estimator will assemble information on the net cost of the works including calculating: the current rates for labour materials and construction equipment, the unit rates, the preliminaries or general items and finally the summaries.

1. Current rates for labour, materials and construction equipment
The rates for labour will be the 'all in' rates based on the basic rates as the national working rule agreement with allowances for labour extras. An example of a detailed build-up of the 'all-in' labour rate is contained in the latest copy of Spon's price book for civil engineering works.[1]

The rates for materials should make due allowance for delivery to site, offloading, storage, unavoidable double handling and waste.

The construction equipment rates should cover for transport to site, erection/dismantling, operators, maintenance and fuel. Major static items of plant such as tower cranes are normally priced separately in the general items or method-related charges section of the B of Q whilst other items are often included in the individual rates.

2. Unit rates for each item in the work sections of the bill of quantities
The three main estimating techniques used by contractors when pricing major construction works are detailed below.
(a) Operational estimating
Operational estimating, which is the recommended method for estimating civil engineering works, requires the estimator to build up the cost of the operation based on first principles, i.e. the total cost of the construction equipment, labour, permanent and temporary materials. The total cost of the operation is then divided by the quantity in the B of Q to arrive at an appropriate rate.

A significant advantage of the system is that it provides a complete integration between the estimate and the programme which in turn enables cash flow forecast to be produced. The process involves:

- compiling a method statement, showing sequence timing and resources required;
- refining the method statement to show an 'earliest completion' programme with no limit on resources;
- adjusting the programme by 'smoothing' or 'levelling' the resources in order to produce the most economic programme to meet the time constraints;
- applying current unit costs: fixed, quantity proportional and time related.

Establishing realistic productivity levels for labour and construction equipment on major operations can prove difficult, particularly in overseas work. However, the operational estimating approach enables the estimating team to appreciate fully the major risks and uncertainties in the works.
(b) Unit rate estimating
Unit rate estimating, which is the standard procedure in the building

sector, involves pricing individual rates in a B of Q which has been prepared in accordance with a method of measurement, e.g. SMM7.

The unit rates are calculated using one of the following methods:

- historical rates based on productivity data from similar projects;
- historical rates based on data in standard price books, e.g. Spon's, Wessex, Laxtons;
- 'built-up' rates from an analysis of labour, materials and construction equipment for each item and costed at current rates.

There are several possible disadvantages of using the unit rate method for estimating major works. The system does not demand an examination of the programme or the method statement and does not encourage an analysis of the real costs and the major risks in undertaking the work. Furthermore, the precision and level of detail in pricing each item can give a false sense of confidence in the resulting estimate.

Generally, it is not recommended that the data from standard price books are used in the estimating of major civil engineering works, either at tender or when quotations for variation orders are required. The reason for this is due to the possible differences in ground conditions, method statements, temporary works, availability of construction equipment, location of the project and the time of year in which the work is to be executed, etc. Each project should be considered on its own merits and the cost estimate based on first principles using the operational method.

(c) Man-hours estimating

Man-hours estimating is most suitable for work which has significant labour content and/or for which extensive reliable productivity data exist for the different trades involved. Typical applications include:

- design work and drawing production, both engineering and architectural;
- installation of process plants and offshore modules.

This method of estimating is frequently used by the major electrical and mechanical contractors as well by the large American contractors, e.g. Bechtel. It should be used in conjunction with a construction programme/schedule in order to highlight any restrictions, e.g. availability of heavy lifting equipment, which may affect the labour hours expended in fabrication yards or on site.

3. General items or preliminaries

Typical general items on a major civil engineering project with an approximate tender value of £40m. (1992 prices) and a 130 week (2.5 years) contract period.

Contractor's site on-costs (time related)

Site staff	Weeks	Rate	Total
Project manager	130	£540	£70 200
2 Agents	2 × 104	£480	£99 840
4 Section engineers	4 × 104	£400	£166 400
4 Engineers	4 × 104	£325	£135 200

2 Engineers	2 × 130	£325	£84 500
4 Engineers (s/c supervision)	4 × 104	£350	£145 600
2 Foremen	2 × 104	£325	£67 600
2 Foremen	2 × 130	£325	£84 500
Quantity surveyor	130	£471	£61 230
Quantity surveyor	104	£325	£33 800
Office manager	130	£385	£50 050
Timekeeper	104	£245	£25 480
Storekeeper/checker	130	£270	£35 100
Storekeeper	104	£270	£28 080
Cost clerk	130	£240	£31 200
Clerk	104	£240	£24 960
Typist/telephonist	130	£155	£20 150
Company cars (senior site staff)	1014	£150	£152 100

Head office staff (time related)

Engineers	130	£450	£58 500
Design engineers	104	£325	£33 800

General overheads
To tender summary £1 408 290

Preliminaries

Description	Labour	Temporary materials	Plant hire	Plant construction	General O/H
Site offices		35 000			
Stores		7 750			
Mess huts	3 250	9 000	2 250		
Toilets		3 250			
Running costs	21 500	8 500			
Plant transport			55 500		
General site labour	107 000				
Electrical connection	4 500	4 500	2 250		
Water connection	1 000	1 000	500		
Haul roads, hoardings	11 000		22 000		
Temporary fencing	2 250	2 250	1 000		

Plant purchases					
Personnel carriers	23 000		28 500		
Vans	23 000		17 250		
Compressors	17 750		26 500		
Pumps	17 750		42 500		
Cranes	40 000		66 000		
Miscellaneous			44 500		
Maintenance	44 500		44 500		
Scaffolding					
Block A	55 500		111 000		
Vertical faces	16 500		22 250		
Around excavations	6 250		7 000		
Access for s/c's	22 250		22 250		
Small tools				44 500	
Plant consumables				389 000	
Contract works insurances (excesses only)					44 500
To tender summary	£417 000	£71 250	£515 750	£433 500	£44 500

Items which should also be considered include: telephones, fax machine, photocopier, print machines, road cleaning, signboards, compounds, security, mobile phones, etc.

4. Preparation of summaries, tender summary, analysis sheets, special conditions

Stage 4 – Commercial appreciation

Following the production of the cost estimate a small management team, comprising the estimating and commercial directors, the chief estimator, the estimator and the proposed contracts manager, will make a separate comprehensive evaluation of the estimate to ensure that the bid is both feasible and commercially competitive.

The first task of the senior management team at this tender committee meeting is to review the estimate taking into account the construction method and programme, the technical and commercial risks, the contract cash flow and finance, the potential for use of own construction equipment, the competition, the economic climate and the commercial opportunities.

On contracts involving major earthworks the risks can be considerable, particularly in connection with borrowpits and quarries as the material to be extracted may subsequently be rejected as unsuitable by the engineer or the local authority may refuse a planning application for extraction.

Weather conditions can also be influential with continuous wet weather likely to cause a prolonged shut-down of all major earthmoving operations, the costs of which may not be recoverable under the contract.

The team will consider the commercial opportunities, particularly the method of billing and whether the major quantities are under- or over-measured or any items omitted entirely, any differences between the specification, drawings or B of Q, the lack of drawings or poor design and the contractor's alternatives.

Stage 5 – Conversion of cost estimate into tender bid

The second task of the senior management team at the tender committee meeting is to convert the estimate into the *tender bid*. The following items are considered and agreed upon:

1. The financial adjustment to be made following the commercial appreciation.
2. The allowances for discounts on subcontractors and suppliers.
3. Late quotations, these could be included as an 'adjustment item' at the end of the B of Q.
4. The contribution for head office overheads – between 4 and 8 per cent.
5. Profit, normally based on what the market can stand.
6. Qualifications to the bid, if any.

At the conclusion of the meeting the estimator will be required to convert the cost estimate to the tender bid, the difference between the two being called the spread. The spread could be added evenly across all the rates, or it could be added to the early work (front-end loading), or it could be inserted as a lump sum in the preliminaries or general items. Whichever method is chosen may have a considerable influence on the profitability of the project depending on the final remeasured quantities.

Stage 6 – Submission of tender

Finally the tender is submitted to the client in the form specified in the invitation letter, arriving at the correct address at the right time. The contractor should keep all copies of the tender documents marking the drawings 'used for tender'.

Following the award a copy of the estimate should be passed to the agent/commercial manager on site for use in establishing the value of variations.

Operational estimating – example

Question

The site for a city-centre redevelopment area requires excavating and the total quantity of the material to be excavated has been calculated to be

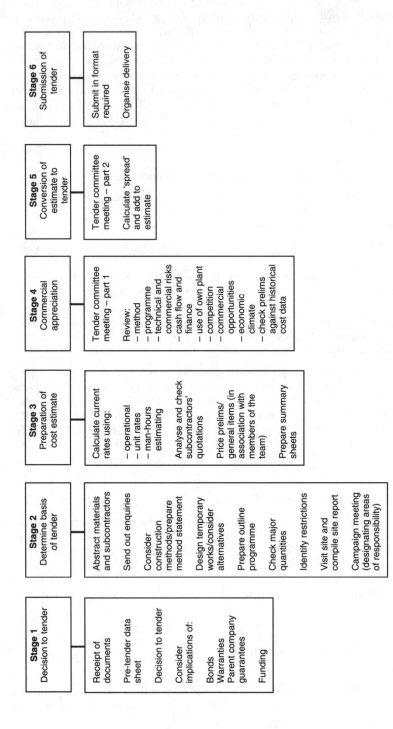

Figure 8.1   Stages in the preparation of the contractor's tender

20 000 m³. The only available tip is 10 miles by road from the site. A contractor has assessed that a tracked excavator, with an output of 60 m³ per hour, will cost £18.50 per hour to hire including the driver. Lorries, with a 10 m³ capacity, will cost £12.50 per hour including the driver.

Using the above information prepare and calculate an 'excavate and dispose' price per cubic metre for the material to be removed. Assume an appropriate figure for profit and overheads and make assumptions about any other costs and factors which you should make allowance for in your calculations. (ICE Examination in Civil Engineering Law and Contract Procedure June 1987, Paper 2)

Possible solution

Assumptions:

1. B of Q project – CESMM3 applies.
2. Size of excavation – 100 × 50 × 4 m deep.
3. Standard 'all-in' rates: unskilled labour £6.50 per hour, ganger £7.00 per hour.

It is also necessary to consider the following factors:

1. The quantity of excavation in the B of Q is the net volume to accommodate the foundations and any structures above them (CESMM3 Class E Rule M5).
2. Any additional excavation required for working space, together with the subsequent backfilling, should be included in the excavation rate (CESMM3 Class E Rule C1) (assume not required in this project due to method of construction – see note 3 below).
3. Any temporary side supports to the excavation should also be included in the excavation rate (assume permanent secant piles installed before basement excavation – so no further supports required).
4. Excavating around existing services and other hand excavation.
5. Temporary works, e.g. dewatering system, if the ground water table is near the surface (assume required).
6. Support of adjacent buildings (assume required).
7. Bulkage factor of excavated material (assume 30 per cent).
8. Temporary ramps and access roads.
9. Cleaning of public highway.
10. Any restrictions on working hours (assume none).
11. Season of the year.
12. Bottoming-up required.

Anticipated duration of excavation is
20 000 m³/60 m³ per hour
    = 333 hours/47.5 hours per week
    = 7 weeks

1. *Excavation* (machine)
   Tracked excavator 333 hours @ £18.50 per hour   6 160.50
   (assume all fuel and fuel distribution
   included in 'General items')
   Earthworks ganger 333 hours @ £7.00 per hour   2 331.00
   Banksman   333 hours @ £6.50 per hour   2 164.50

2. *Excavation* (hand)
   Estimate 50 hours for a gang of three
   Labourer   150 hours @ £6.50 per hour   975.00

3. *Dewatering system*
   Mobilise construction equipment   £500.00
   Install 20 no. wellpoints @ £250 each   £5,000.00
   Demobilise on completion   £500.00
   2 no. 150 mm pumps and hoses
   6 weeks × 2 × £400 per week   £4800.00   10 800.00
                                          _____

4. *Support of adjacent buildings*
   Temporary support system to adjacent shops
   (based on subcontractor's quotation)   5 000.00

5. *Temporary ramps/access roads*
   Provide/maintain and clear off   1 500.00

6. *Disposal*
   Assume bulking factor 30 per cent
   (depends on type of ground – refer to soils
   investigation reports)

   Output in solid 60 m³ per hour
   Actual output (bulked) = 60 m³ × 1.3 = 78 m³ per hour

   Lorry cycle:

   Loading 10 m³/78 m³ per hour × 60 min.   = 8 min
   Lorry to tip 10 miles/15 mph.   = 40 min
   Offload at tip   = 5 min
   Lorry to site 10 miles/20 mph.   = 30 min
                                                _____
   Lorry cycle   = 83 min
                                                _____

Lorries required to move 78 m³ per hour, so
78 m³ (loose)/10 m³ (loose) = 7.8 loads per hour

1 lorry does
60 min/83 min = 0.72 complete journeys per hour

No. of lorries required is
7.8 loads per hour/0.72 = 10.8   (say 11)

11 No. lorries × 333 hours @ £12.50	45 787.50

Tipping charges:

20 000 m³ × 1.3/10 m³ = 2600 loads @ £4.00	10 400.00

Cleaning of public highway:

Tractor and brush	
100 hours @ £10.00 per hour (including driver)	1 000.00
3 labourers, 300 hours total @ £6.50 per hour	1 950.00
Estimate at cost	£88 068.50
Add site overheads 7 per cent	£6 164.80
	£94 233.30
Add head office overheads and profit 8 per cent	£7 538.66
Total price to client	£101 771.96

Price per m³ of excavate and dispose for B of Q is
Total price/net volume = £101 771.96/20 000 m³ = £5.09 per m³

Tutorial questions

1. What are the main elements of cost that should be taken into account when building up a tender rate? How would an estimator assess each element?

    Suggest how you would attempt to improve cash flow on a contract involving substantial temporary works at an early stage and the use of expensive specialist plant (e.g. a travelling shutter or launching gantry) measured in accordance with CESMM.

    A project for which you are preparing a tender requires considerable deep excavation for which you propose allowing for the use of temporary sheet piling. However, you believe that battered side slopes may be a suitable alternative at a similar cost. How would you insert the money necessary for this work into the tender bills? (ICE Examination in Civil Engineering Law and Contract Procedure, June 1983, Paper 3)

2. You are a contractor's estimator and your firm has been invited to tender for the complete replacement of the existing lighting in a road tunnel complex, total length 6 km.

   The tunnel is constructed of reinforced concrete with steel internal linings, roughly semicircular in cross-section above the roadway. It carries four lanes of traffic and is normally in use 24 hours per day, 7 days per week. There is one main entrance and one branch tunnel at each end. The branch tunnels are 1000 m long.

   The contractor will be permitted to utilise one of the branch tunnels as a site compound as this will be closed for the contract duration. The main tunnel will be closed between the hours of 10.00 pm and 6.00 am. No work on site will be permitted outside these except in the area of the site compound.

   During the working period there is an obligation to leave one lane open for emergency services. Each night's work station must be left clean for the following day.

   The new lighting is to be installed centrally in the tunnel throughout its length at a height of 10 m above the road level. Ceiling height is generally 11 m above road level except in the position of two ventilation shafts where the ceiling level drops to 9 m above the road level.

   The work consists of drilling the ceiling of the tunnel at 6 m centres for fixings, fabrication and erection of structural steel support systems and the installation of the light fittings and wiring. The fittings are a continuous run of fluorescent type fittings. The existing lighting installation does not interfere with the new lighting position and accordingly needs only to be removed on completion of the new installation.

   The contract duration is 26 weeks which necessitates simultaneous working on a number of fronts. To be competitive, however, the number of general purpose vehicles and drivers must be kept to a minimum.

   (a) Describe your approach to the problem of achieving maximum hours at the work face. In particular, consider how you would deal with the specific problems of:
      (i)   access/scaffolding
      (ii)   canteen/toilet facilities
      (iii)   protection and site clearance
      (iv)   site communication
      (v)   site supervision
   (b) Show by calculation the actual number of working hours at the work face, per man, per shift, you would allow in your tender

3. With regard to the specific problem outlined in question 2, schedule the basic information you would give in enquiries being sent out to:
   (a) suppliers
   (b) subcontractors
   Define, in addition, the subcontractor's responsibilities which would be covered in the enquiry

4. You are a specialist sub-contractor tendering to become the nominated electrical sub-contractor for the tunnel lighting. Detail those items which you consider should be covered in your pre-tender site visit report, bearing in mind that this project is in a geographical region not normally covered by your organisation. (Questions 2, 3 and 4, RICS Direct Membership Examination, Tendering and Pricing, March 1985)

5. Describe the steps you would take as a contractor to ensure that before submitting a tender you had obtained for yourself all the necessary information as to site conditions, risks, contingencies and all other circumstances influencing or affecting your tender, as required in Clause 11 of the ICE Conditions of Contract (5th edition) (ICE Examination in Civil Engineering Law and Contract Procedure, June 1985, Paper 3)

Reference

1. Davis Langdon and Everest (eds) 1992 *Spon's Civil Engineering and Highway Works Price Book* 6th edn E & F N Spon

Further reading

CTA Services (Swindon) Ltd 1990 Distance Learning Course, Civil Engineering and Contract Procedure

Hudson E 1988 Contractor's tenders – the tender approach. *Chartered Quantity Surveyor* June: 17

Lecture given by Andrew Norman, Consultant, Advanced Project Management course, UMIST, 1992

Lecture given by Ron Steel (former chief estimator of Balfour Beatty Construction (Northern)), MSc Construction Management course, University of Loughbourgh, 1985/86

McCaffer R, Baldwin A N 1991 *Estimating and Tendering for Civil Engineering Works* 2nd edn BSP Professional Books

9

Construction programming

Introduction

The contract administrator and the commercial manager should have a full understanding of the programmes and programming methods used on construction projects. The programme will be of particular importance when:

1. Establishing the contractor's estimate – identification of time and cost of general items/preliminaries (on-site overheads and head office costs) and time required for alternative construction methods.
2. Compiling and monitoring a cash flow forecast for the employer or the contractor.
3. Monitoring the contractor's progress on site.
4. Assessing the value of interim payments including preliminaries or general items.
5. Establishing the extension of time and the associated cost to be awarded for delays and variations.

The programme, or as the Americans call it 'the schedule', provides the project manager with the main means to plan and control the works. The programme is not a document to be pinned on the site office wall on day one and forgotten. It should identify the most efficient way of carrying out the work and should show the priorities, it should be flexible and simple to follow and should be updated regularly and quickly to reflect the changing situation on the project.

The programme should not be drawn up in isolation by the contractor's central planning department and then imposed on the site organisation; it should be formulated, agreed and monitored by the construction team who will be responsible for its execution. Indeed one of the unwritten rules of sound construction management is 'a project can only be managed on the site itself'.

Furthermore it should be appreciated by all parties that the programme is at best an approximation of intent. Variable factors on any complex construction project can be of such magnitude that the tender programme may not truly reflect a realistic assessment of the situation. Such factors might include efficiency and performance of: contractor's site management, labour force, construction equipment, subcontractors and suppliers, weather, ground conditions and last but not least the performance of the client and his representatives.

Compiling a programme

The construction programme can be compiled only after full consideration of the construction techniques and methods proposed have been identified, i.e. after completion of the construction planning stage. The programmer will need to consider and choose the most appropriate programming technique after identifying the status and level of detail required.

Further items which will need consideration by the programmer include:

- critical dates for completion and 'milestone' dates;
- the resources, i.e. labour, construction equipment and materials reflecting the construction techniques and methods proposed;
- the activities, i.e. packages of work which consume resources;
- the duration of each activity which depends on the level of resources, the production rates and the quantity of work to be completed;
- total demand for resources;
- the logic or construction sequence, i.e. the relationship and inter-dependence between the activities;
- the critical path;
- the working dates in the calendar and potential problems, e.g. seasonal working;
- external restraints which may also affect the duration such as the delivery date of key items of equipment or delayed access to the part of the works;
- types of labour (labour-only subcontract or direct labour);
- subcontractors (nominated or domestic);
- holidays, shut-down periods (e.g. maintenance weekends);
- hours of working (e.g. earthmoving fleet – 7 day/90 hour week, joiners – 40 hour week);
- other contractors on site (e.g. power station project – boiler, structural steelwork, chimney and cooling towers contractors);
- client's financial restraints (particularly in developing countries).

Types of programme

Bar charts

The bar chart or Gantt chart is easy to understand and as a result is the most common form of programme encountered on all but the most complex projects. The bar chart readily shows the timing and durations of activities but does not clearly show the relationship between them.

Bar charts are used in the following cases:

- master programme on simple projects (short time-scale/few activities);
- short-term programme (weekly – tied to master programme);

- master programme on complex jobs (translation from network for non-specialist uses);
- tender programme, often with staff and major items of construction equipment shown.

Bar charts are extremely versatile and can easily be adapted depending on the priority; the space within the bars can be used for information on resources, for example number of key tradesmen and the space below the bars can be used for the actual monitoring of progress (different colours or shading can be used for different months).

Alternatively progress could be monitored using the PSA *Planned Progress Monitoring* method of counting activity weeks achieved compared to the programmed activity weeks achieved at any point in time.[1]

Unfortunately bar charts can be misleading in reflecting too simplistic a picture. They do not, for example, reflect the complex interrelationships and dependencies between the activities to be found on most construction projects and by themselves do not lead to a high degree of control.

Linked bar charts

Linked bar charts (Fig. 9.1) are a development of the Gantt chart – retaining the visual simplicity whilst at the same time emphasising the logic. Dependencies between activities are represented by vertical links between the completion of one activity and the start of another. However, on large bar charts for complex projects it is very difficult to show all the interrelationship links as the chart becomes visually 'over-detailed'.

Line of balance

The line of balance or elemental trend analysis method of programming is appropriate only for repetitive projects, e.g. new housing schemes, refurbishment of houses and office blocks, long bridge decks, multi-storey structures. The advantage of using this method is that it does enable the securing of maximum utilisation of resources.

The time-scale is identified on the horizontal axis while the vertical scale represents the completed units. An inclined line is drawn representing the estimated timing and location of each separate trade throughout the project (see Fig. 9.2).

The gradient of the line reflects the anticipated production rates which are dependent on the level of resources. An increase in resources will mean completion in shorter time reflected in a steeper angle on the line. A buffer zone between the lines represents relationships between the trades.

The graph can be monitored and redrawn to represent actual productivity of the trade gangs and corrective action taken or a revised final completion date established.

The line of balance method provides a good means of control but is not suitable for use on non-repetitive work.

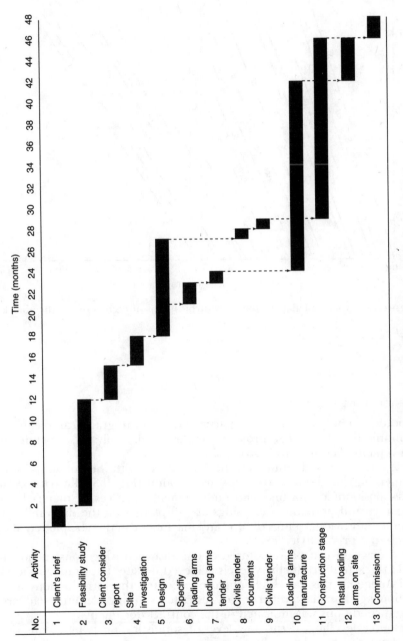

Figure 9.1   Linked bar chart – oil jetties project

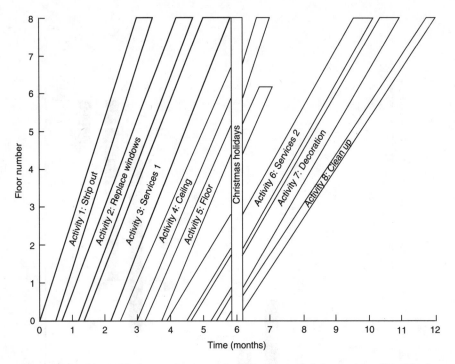

Figure 9.2   Line of balance chart – refurbishment of eight-storey office block

Location–time diagram

A location–time diagram is a particularly useful graphical means of communication for linear projects such as roads, railways, pipelines or even repetitive multi-storey projects.

The time-scale is identified on the vertical axis with the horizontal scale representing the distance from a starting-point (Fig. 9.3). Restrictions in access, positions of cuttings and embankments, sources of material and temporary and permanent crossings are all plotted on the diagram. The location of individual gangs of men and machines can then be plotted after considering productivity levels.

The diagram can be monitored and redrawn during construction to reflect the actual productivity levels achieved on site. If production levels are inadequate this can highlight the need for the introduction of an additional gang or be a reflection of restricted working conditions/adverse ground conditions not indicated in the contract documents.

Resource levelling at this low level of sophistication is by means of a histogram and bar chart adjustment and not by any complex calculation on the network itself.

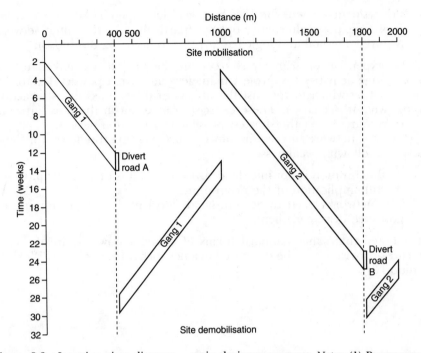

Figure 9.3   Location–time diagram – main drainage contract. *Notes*: (1) Programme based on two gangs each working at 40m per week. (2) 2-week cycle for: excavation, trench support, pipe bed, lay sewer, backfill. (3) 1 week allowed for remobilising gang at new location

Network analysis

Networks comprise a family of techniques as shown in Fig. 9.4. Programmes compiled based on networks show the logic and the interaction between activities and have certain strengths when compared to bar charts including:

1. The shortest duration for the project can be calculated – i.e. the critical path can be identified.
2. The effects of potential problems and delay can easily be identified.

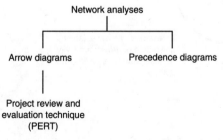

Figure 9.4   Network analyses

3. Project management software has been developed based on networks.
4. It often exposes alternative options (particularly as the initial network is purely method and logic based prior to inserting durations).

If the expensive or scarce resources on the project are linked to the programme activities the project manager will have a powerful management tool by which to forecast the effects of different courses of action using 'what-if' scenarios. These exercises can result in the most efficient use of resources over the shortest possible construction period.

However, networks do require more effort and expertise to compile and have the following weaknesses:

1. As the duration is not initially drawn to scale it is often difficult to see the full implication of the programme.
2. The networks need to be translated into bar chart form for use by non-specialists, e.g. clients.

It is noted that some standard forms of contract now require the contractor's programme to be submitted in a network format (e.g. GC/Works/1 (edition 3)).

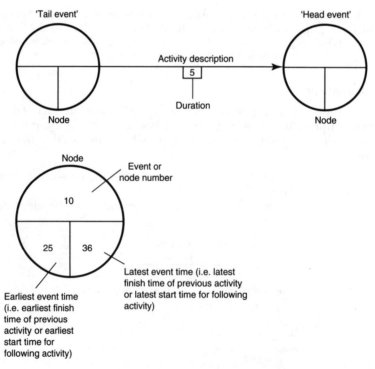

Figure 9.5   Arrow diagram – convention

Activity-on-arrow networks

The arrow diagram (Fig. 9.5) is made up of two basic elements:

1. Arrows which represent activities, or tasks, which is the specific work or job which takes place over a period of time.
2. Events, or nodes, which are instantaneous moments in time occurring between one activity finishing and another starting.

The activities need not be real, in that no energy is consumed e.g. waiting for concrete to cure, but there must be some duration in hours, days or weeks.

'Dummy activities', which have no resource or duration, are normally shown as a dotted arrow and may be necessary to show restraints within a programme, to maintain logic, to ensure unique activity numbering, or to avoid dangling events (Fig. 9.6).

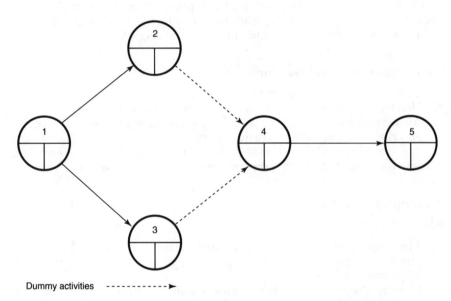

Figure 9.6 Arrow diagram. Dummy activities have no duration – they maintain the correct logic within the diagram and ensure unique activity referencing. Activities 1–2 and 1–3 are progressing concurrently and both must be complete before activity 4–5 may commence

Fundamental guidelines for producing networks

It is necessary to ask three questions:

1. What activity must immediately precede this operation?
2. What activity can immediately follow this operation?
3. What activities can be taking place concurrently with this operation?

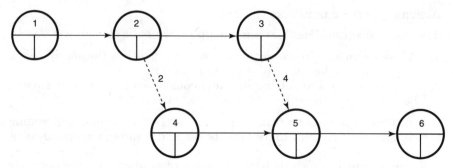

Figure 9.7 Ladder diagram. Activity 4–5 may commence 2 days after activity 2–3 has commenced, i.e. this is known as 2 days' lead time. The earliest date for completion of activity 4–5 is 4 days after the completion of activity 2–3, i.e. this is known as 4 days' lag time

Generally no activity may start until the previous activities in the same chain are complete. However, lead and lag times can be accommodated within arrow diagrams by formulating 'ladders' (Fig. 9.7).

Conventions of arrow diagrams

1. Time flows from left to right.
2. Head events always have a higher number than tail events.
3. All events, except the first and last, must have at least one activity entering and one activity leaving them.
4. All activities must start and finish with an event (however some activities can finish with a dummy arrow leading to another event).

Drawing up an arrow diagram

Basic points:

1. The logic and durations in network diagrams should be drafted manually even if using project planning software; drawing the plan helps you think ahead.
2. Identify the major events/activities – some intermediate events may require an imposed date.
3. For every activity ask: what has to be done before this and what can be done now?
4. Check that the network contains no:
   (a) 'loops' (a path through the network which loops back on itself and therefore cannot be analysed), or
   (b) 'dangles' (an activity which is linked to the network by the tail event but not by the head event).
5. Review the diagram remembering:
   (a) arrow lengths are not significant;
   (b) events can be separated by dummy activities to increase clarity;
   (c) you will probably have to draw it again so do not make it a work of art.

Calculation of the total project time

The total project time is the shortest time in which the project can be completed using the sequence and the longest path through the duration of activities as shown in the arrow diagram. The sequence(s) of events constituting this time span is called the *critical path*.

The first event (the one to the extreme left), should be given the time zero. Then proceed to each event in order and calculate the earliest time at which each event can occur. If several activities lead into one event the earliest date for the event is fixed by the longest chain leading into it. This process should be repeated until the earliest date is fixed to the final event – the whole process being known as the 'forward pass'.

The 'critical path' is identified by reversing the process by performing the 'backward pass'. Starting at the final event give this event its earliest completion time as identified above. By subtracting duration times, calculate the latest possible time for each event. As before, the latest date for an event is fixed by the longest chain leading to the event but in this case working backwards in time. When the first event is reached you should have reached the time zero.

Assuming that no mistakes have been made the critical path will be found from that chain (or chains) of events where the two times are identical and where the activity durations in the chain are equal to the differences between the times of each pair of 'head' and 'tail' events. (Remember: take the latest date on forward pass and the earliest date on backward pass.)

Example No. 1

Figure 9.8 shows the durations (in weeks) and the interdependencies of five activities required to carry out a project:

1. Construct an arrow diagram assuming no resource restrictions.
2. Calculate the minimum duration of the project and show the critical path.

Figure 9.9 gives the solution to Example No. 1, where the critical path is shown thus // and the minimum duration of the project is 23 weeks.

It is possible to have more than one critical path running through a project. However, in the case of Example No. 1 the critical path could be identified by inspection or by applying the second test.

Consider the activity in question: subtract the earliest event time at the tail of the activity from the latest event time at the head of the activity and then subtract from the result the duration assigned for that activity. If the result is zero then the arrow is on the critical path.

e.g. activity $2-5$: $23-5 = 18-10 = 8$
thus activity $2-5$ is *not* on the critical path.

In certain circumstances it can be extremely difficult to estimate the duration of some types of work; for example, nuclear power station construction often involves new technology which gives rise to new construction techniques never before used. In this situation it is necessary to

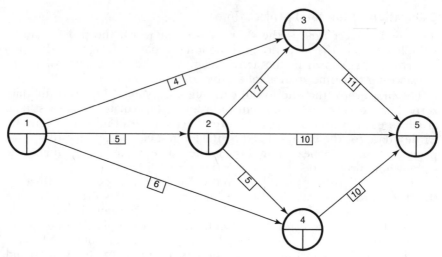

Figure 9.8   Example No. 1 – simple project

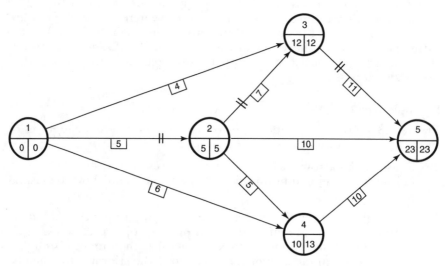

Figure 9.9   Example No. 1 – simple project – solution

make probabilistic analysis of durations and combine such analysis with normal activity on the arrow networks. Such a system, called project review and evaluation technique (PERT), is beyond the scope of this chapter; see Pilcher's *Principles of Construction Management* for full details.[2]

Precedence diagrams (sometimes called 'activity-on-node')

A graphical representation of a precedence diagram resembles a flow chart. The system lends itself better for use with computer software project planning packages and has the following advantages over an activity-on-arrow network:

1. The logic is defined in two stages giving greater flexibility.
2. Dummy activities are eliminated.
3. Revision and introduction of new activities is simple.
4. Overlapping of activities is more easily defined.
5. The analysis of float time available to non-critical activities is much easier.

However, with precedence diagrams it is more difficult to follow the flow of work as shown on the programme drawing; as a result it is easier for logic errors to arise, particularly on a large complex project.

Activities are represented by 'boxes' or 'egg-shaped nodes' (Fig. 9.10) and the interrelationships between activities by lines known as dependencies.

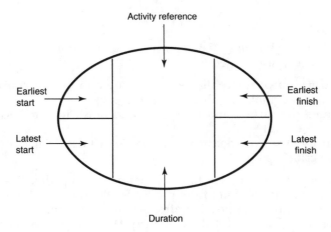

Figure 9.10   Precedence diagram – egg-shaped node

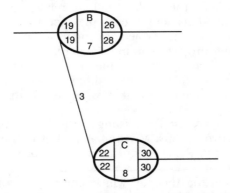

Figure 9.11   Precedence diagram – overlapping activities. Activity C may commence 3 weeks after B has commenced

Table 9.1   Example No. 2

Activity	Duration (days)	Depends only on completion of:
A	7	—
B	9	A
C	11	A
D	4	B and C
E	7	C
F	9	C
G	11	E and D
H	4	B with 7 days' lead time
J	6	E and F
K	2	H and G and J

Example No. 2

Table 9.1 shows the interdependencies and durations of ten activities required for a typical project.

1. Construct a precedence diagram assuming no resource restrictions.
2. Calculate the minimum duration of the project.
3. Schedule the earliest and latest start and finish for each activity and show the critical path.
4. If activity D is extended to a duration of 10 days, what is the effect on the critical path?

The solution is shown in Fig. 9.12, where the critical path is A–C–E–G–K and the minimum duration of the project is 38 days.

Figure 9.13 shows the position after activity D is completed in 10 days instead of the anticipated 4 days. Activity D is now on the critical path A,C,D,G and K. The 3 days of float are used up and the project will be delayed by 3 days finishing on day 41.

The completion may in fact be delayed for longer than 3 calendar days depending where the 3 days fall in relation to weekends/Bank Holidays/shut-downs. This simple example therefore highlights the importance of identifying the project calendar – one of the first steps executed when using project planning software.

Establishing the 'float' or spare time in the activities is particularly important as this spare time will be utilised to adjust the timing of the activities in order to obtain the best possible use of resources (known as resource levelling).

It is usually necessary to make a listing of the various properties of each activity within a network, thus: activity (tail and head numbers), duration, earliest start, latest start, earliest finish, latest finish, total float, free float and independent float.

The 'total float' is the time by which any activity may be extended without interfering with the project end date, i.e. the latest time of the finish event less the earliest time of the start event less the duration.

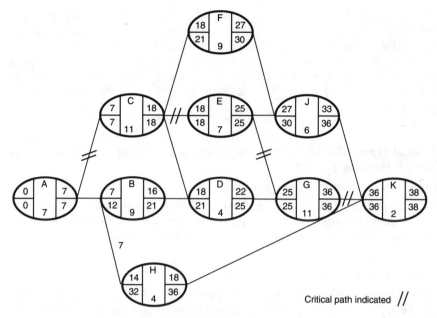

Figure 9.12   Solution to Example No. 2(a)–(c)

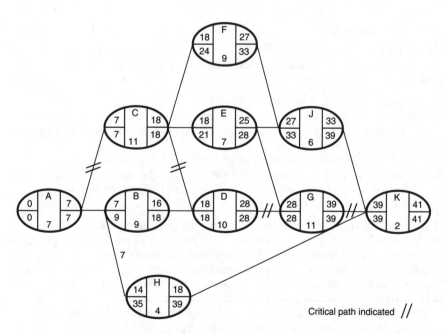

Figure 9.13   Solution to Example No. 2(d)

The 'free float' is that time by which an activity can be extended without affecting subsequent activities, i.e. the earliest time of the finish event less the earliest time of the start event less the duration.

The 'independent float' is that time by which an activity may be extended without affecting any other activity, either previous or subsequent, i.e. the earliest time of the finish event less the latest time of the start event less the duration.

Typical types of programmes required on the Hong Kong Mass Transit Railway project

Engineer's preliminary programme

This is produced by the client's PM team and/or the design consultants. It indicates the engineer's assumed method of construction and is used as an aid to fixing the original critical dates and milestones.

Tender programme

This is submitted by tenderers as part of the tender documents. It is unlikely that this programme goes into much detail and many are simply variations of the preliminary programme. This programme is particularly important if the tenderer is proposing an alternative construction method or sequence as part of the tender. The target completion dates stated by the contractor may be earlier than the contract completion dates.

Contract programme

Required under the equivalent of clause 14 of the FIDIC 4th and the ICE 6th and must be submitted in accordance with the requirements of the general specification, i.e.

1. Must be submitted within 60 days.
2. Must be in network form with network analysis showing earliest and latest start dates for each activity, activity duration and floats against critical dates and milestones.
3. Must have no activity duration greater than 12 weeks.
4. Must meet critical dates, milestones and take account of any programming restraints mentioned in the particular specification.
5. Must take account of the activities of interfacing contracts.

This programme is considered of prime importance as it is a statement of the contractor's intention at the time of the award. The programme is thoroughly examined by the engineer to see whether it meets the requirement of the specification, whether it represents a reasonable and logical sequence of work, that durations for activities are realistic and achievable and that any indicated restraints are valid.

The engineer makes formal comment on the programme following his analysis; these comments should be incorporated into the programme and a revised programme submitted. When everyone is satisfied the programme is accepted as the 'contract programme'. Just as the B of Q is used as the basis for the assessment of variations and additional works so the contract programme is the reference against which 'time-related elements' are assessed.

Working programmes

As time progresses during the construction phase the contract programme may lose its relevance to the physical work actually carried out on site. For example, the progress on site may not conform with the approved programme, or the contractor may change the methods of construction. In these cases a revised working programme indicating how the contractor intends to meet the completion date for the project is required. The general requirements for the form and content should be the same as required for the contract programme.

Three-monthly programmes

The general specification requires the contractor to submit a 3-month bar chart on a 'rolling' basis at monthly intervals. These bar charts should show in greater detail the activities in the latest comprehensive programme.

Engineer's combined installation programme

This programme, produced by the client's PM team, covers the critical period from main structure completion to contract completion. The programme attempts to co-ordinate the activities of all the various designated electrical and mechanical contractors appointed under direct contracts with the client, with those of the civils contractors.

The preliminary version of the programme is issued to contractors for comment and after minor revisions reissued. The programme defines the periods when the designated contractors are to be expected to carry out their works in a particular area and enables the civils contractor to plan the finishing works more efficiently.

Effective programming in practice

The case study considers the construction of an eight-level deep basement underground station in Hong Kong: the method statement indicated a 'top–down' construction technique. The study briefly demonstrates how the client's PM team, by taking a pro-active as opposed to a passive role, can help a contractor to achieve early completion and thus benefit both parties to the contract.

The particular contract required at least twice the programming input of any other project on the same section. The client's PM team soon realised, even before the work commenced on site, that the contractor had been given an almost impossible task to complete the basic structure by the specified date, not allowing for any late design changes.

Following a realistic appraisal of the contractor's programme the client's

PM team recommended a number of relaxations in the specified engineer's requirement including the public closure of a critical access road. A further idea suggested by the client's team was a modification to the contractor's method statement. The team suggested sinking shafts to the bottom of the eight-level basement in order to enable the rock to the lower three levels to be removed in caverns before the 'top–down' construction reached the 'rock-soft' interface.

These positive suggestions by the client's PM team together with excellent production performance achieved by the contractor enabled the basic structure to be completed 8 weeks early.

Additional difficulties were encountered on this project including encountering unforeseen ground conditions (old concrete caissons and piles) and late major design changes. The various changes to the works provoked a number of claims for extensions of time and additional costs.

The nature of the work resulted in a stream of revisions to the original clause 14 programme. As these revised programmes were received the contractor submitted proposals to revise the payment schedules so as to protect cash flow; this was generally supported by the client's PM team. With the goodwill of both parties the project was completed in time to the satisfaction of the engineer.

Tutorial question

The author is grateful to Professor F. C. Harris for permission to publish this case study.

Construction of an earth fill dam for a new reservoir

Description of the works

In order to provide drinking water for a nearby population an earth dam is to be constructed to form a reservoir of 1 000 000 m³ storage capacity (Fig. 9.14).

The dam is to be constructed from compacted earth obtained locally and from a quarry upstream. The height to the crest is approximately 20 m and length 100 m. The slopes batter at 1 : 1.75 and are to be rounded into the valley sides in keeping with the general contours of the landscape.

The downstream slope is to be covered with a 300 mm deep layer of loam dressed with 200 mm of topsoil and grass seeded. On the upstream side an abutment (Fig. 9.15) is provided along the base of the dam, and the dam slopes are sealed with a bed of hardcore, two layers of 60 mm thick asphalt and finally a single coat of sealer asphalt.

The boundary between the concrete abutment and asphalt is sealed with a layer of copper sheeting (Fig. 9.16).

The dam is founded on a bed of hardcore overlying the natural rock surface of the valley. A system of pipes is installed into this bedrock to drain off any ground water likely to build up when impounding is complete.

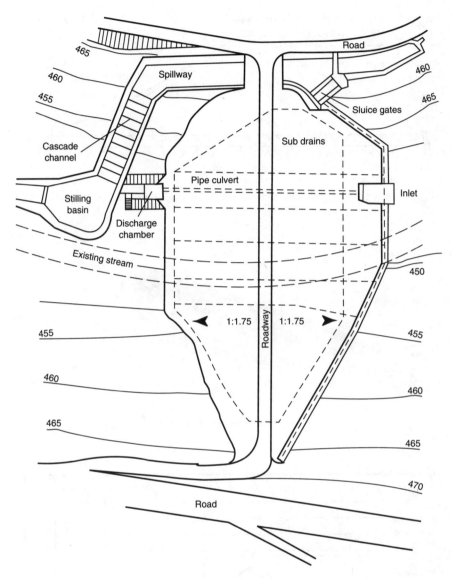

Figure 9.14   Plan of dam

A concrete culvert passing through the body of the dam holds an 800 mm diameter steel pipe to carry away the water from the drainage system and two 300 mm diameter pipes to carry water to the filter plant (Fig. 9.17).

During the construction phase, the culvert is to act as a temporary channel for the existing stream flowing down the valley. A concrete spillway is to be constructed at the north end of the dam with the steel

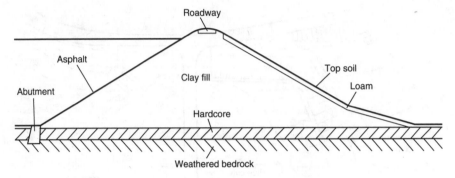

Figure 9.15   Section through dam

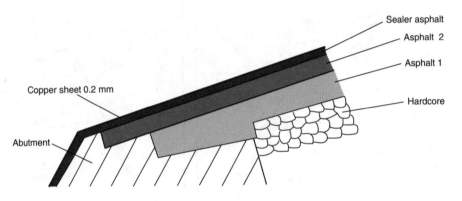

Figure 9.16   Seal between dam and abutment

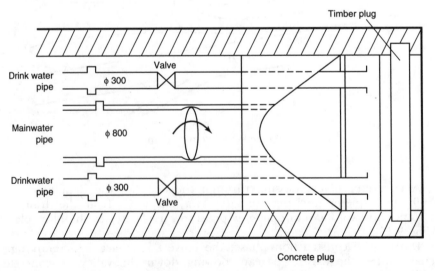

Figure 9.17   Plan through culvert at the inlet

sluice gates to control the height of the water in the reservoir. Any excess water is allowed to cascade down a steep channel into a stilling basin before being allowed to re-enter the flow downstream.

1. Prepare a precedence network analysis in accordance with the attached method statement and then calculate:
   (a) the critical path for the project;
   (b) the earliest and latest times together with the total float for each event.
2. By consideration of the resources described in the method statement, interpret the network in the form of a linked bar chart and reschedule the activities to comply with the resource limitations including the restraint that the dam must be sufficiently complete 300 days after commencement of the construction work to allow impounding of the water.
3. Working in groups select a reasonably representative network of no more than 50 activities and analyse it using a computer software package. The computer analysis should be for time analysis and resource analysis. It should be based on time periods (weeks or days) and not on calendar dates.
4. Working as individuals give your views, in no more than 500 words, on the advantages and disadvantages of using computer software as an aid in the planning and control of construction projects.

Method statement

Main contract – construction of all earthworks, reinforced concrete and other concrete works involving the sealing of the dam surface.

Subcontracts – injection grouting to seal the bedrock, supply and installation of sluice gates, asphalt surfacing of embankments, installation of all steel pipework.

Resources available

1 No. general gang, 2 No. gangs for concrete, formwork and fixing reinforcement, 1 No. central batching plant with fixed position concrete pump, 3 No. bulldozers, 1 No. crawler dragline excavator, 1 No. crawler backacter excavator, 1 No. rear dump truck, 1 No. pulled vibrator roller, 1 No. mixer plant for asphalt production, 1 No. asphalt paving machine, 1 No. compressors, 1 No. crawler crane, 1 No. tracked loader shovel. Also physical restraints.

1. Before placing of the dam fill may begin the following activities must be complete:
   (a) topsoil removal;
   (b) drainage system and injection grouting;
   (c) abutment;
   (d) culvert, discharge chamber and the inlet through which the existing stream will be temporarily diverted until the water impounding begins;
   (e) stilling basin.

2. Placing of loam and topsoil and the downstream face and the water-proofing operations on the underside cannot be started until all the fill is placed.
3. As soon as sufficient of the upstream waterproofing is complete the inlet may be temporarily plugged and impounding commenced. The necessary culvert pipework can then be installed in the dry.

Construction procedure

Preparatory work

Construction work begins with setting up the offices and site yard, surveying and clearing the dam site. The topsoil is stripped by bulldozer and dozed to a spoil heap in the immediate area downstream. Meanwhile the dragline excavates the foundations of the culvert, discharge chamber and stilling basin, during which time a hydraulic backhoe excavator excavates the upstream inlet and the abutment excluding the section through which the existing stream is flowing.

During this phase the two concrete work gangs erect the concrete batching plant and concrete pump and then progressively complete the concreting works using the crawler crane to handle formwork and reinforcement.

With the culvert and associated works complete the stream is diverted and the remaining section of abutment constructed. The dragline is temporarily diverted from work on the installation of the drainage system to carry out these excavation works. The drain trenches are broken out with jackhammers using the general gang and the compressor. The hydraulic backhoe works on the excavation for the spillway after completion of its duties on the abutment and inlet.

Earthworks

Placing of fill for the dam may begin when 50 per cent of the drainage system is installed. The fill material is obtained from the spillway excavation and from a nearby quarry. The quarry utilises the bulldozer, now fitted with a ripper attachment, a large tracked loader and a rear dumptruck. At the dam site two bulldozers and a vibrator roller are needed.

Waterproofing

With the fill operations complete all necessary plant may be sent off site, with the exception of a bulldozer used to spread topsoil on the downstream face and finally to prepare the road which will run across the top of the dam. Meanwhile the subcontractor carries out the waterproofing on the upstream face and finally lays the bituminous road wearing surface.

Spillway

Independent of progress made on the dam itself construction of the spillway may go ahead using the workgang No. 2 in the order: foundation and base for spillway crown, concrete work for sluice gates, wing walls, bridge and part of the cascade channel. Gang No. 1 constructs the stepped section of the cascading channel and control house on the downstream

Table 9.2   Schedule of activities and resources

Activities	Duration (days)	Resources
Excavation work		
Prepare site	2	Bulldozer, general gang
Set up site	5	General gang
Strip topsoil	10	Bulldozer
Put in dam drainage	50	Dragline, general gang, compressor
Divert stream	1	Dragline
Excavate abutment and inlet	20	Backacter, general gang, compressor
Excavate abutment near stream	5	Dragline, general gang, compressor
Excavate culvert	15	Dragline, general gang, compressor
Excavate discharge chamber	5	Dragline, general gang, compressor
Excavate stilling basin	15	Dragline, general gang, compressor
Excavate spillway	20	Backacter, general gang, compressor
Excavate cascade channel	30	Backacter, general gang, compressor
Place fill to dam	65	3 No. bulldozers, vibrator roller, dumptruck, general gang, loader shovel
Place loam and dress downstream face of dam	20	Bulldozer, vibrator roller, general gang
Construct road on crown of dam	15	Bulldozer, vibrator roller, general gang
Reinstate quarry	2	Bulldozer, vibrator roller, general gang
Concreting work		
Set up concrete batching plant	10	2 concrete gangs, crawler crane
Construct abutment	35	2 concrete gangs, crawler crane, concrete pump
Construct culvert, inlet and discharge chamber	80	Concrete gang No. 1, crawler crane, concrete pump
Construct stilling basin	25	Concrete gang No. 1, crawler crane, concrete pump
Construct spillway	80	Concrete gang No. 2, crawler crane, concrete pump
Construct cascade channel and pump house	50	Concrete gang No. 2, crawler crane, concrete pump
Subcontract and miscellaneous		
Injection grouting	20	Subcontract
Weatherproofing to Upstream face of dam	40	Subcontract, bitumen batching plant, bitumen paver
Fix temporary plug upstream in culvert	1	General gang, crawler crane
Erect pipework in culvert and pump house	5	Subcontract

Table 9.2 (cont)

Activities	Duration (days)	Resources
Erect screens in inlet	2	Subcontract, crawler crane
Place concrete plug in culvert and hardening	5	Concrete gang No. 1, concrete pump
Erect sluice gates	12	Subcontract
Inspection and testing	2	General gang
Clear site	5	General gang

side. Both gangs and construction equipment may then be sent off site and the subcontractors allowed to erect the pipework.

Impounding
The inlet area to the culvert is so constructed that heavy timbers may be slotted in to seal off the stream and allow a concrete plug to be placed *in situ*. The pipework, valves and penstock can then be fitted. The removal of the timber seal requires the aid of a driver as the water is continuously impounded during this period.

Site clearing and reinstatement of the landscape disturbed by the quarry activities may take place during the latter phase before the last bulldozer is sent off site.

References

1. Department of the Environment, Property Services Agency, 1986 *Planned Progress Monitoring* Professional Practice & Management Group, Directorate of Quantity Surveying Services
2. Pilcher R 1992 *Principles of Construction Management* 3rd edn McGraw-Hill

Further reading

CIOB *Programmes in Construction . . . a Guide to Good Practice*
Cooke B 1992 *Contract Planning and Contract Procedures* 3rd edn Macmillan
Cooke B 1988 *Contract Planning Case Studies* Macmillan
Harris F, McCaffer R 1989 *Modern Construction Management* 3rd edn BSP Professional Books
Milligan R A 1989 Planning the planning. In Wearne 5 (ed) *Control of Engineering Projects* Thomas Telford ch. 3
Neale R H, Neale D E 1989 *Construction Planning* Thomas Telford
Pearson W Programming and critical path methods, unpublished lecture notes (HKMTR)
Reiss G 1992 *Project Management Demistified* E & F N Spon
Smith N J Advanced Project Management course notes, UMIST
Trimble G Project Control Early Papers, University of Technology, Loughborough

Section E

Management of the post-contract stage

10

Cost control and monitoring procedures

Introduction

Cost control should be carried out throughout the life cycle of the project from inception until settlement of the final account. The implemented system should require that no one makes decisions to commit resources and expenditure without first considering the full consequences of their actions. Furthermore, it should not only record the past but should look forward to predict the anticipated final cost of the project.

At the design stage prior to tender the client's cost control system should ensure that the tender is secured within the estimated budget. This will require a constant checking and rechecking of the project cost model upon production of the design drawings.

Up until the acceptance of the contractor's tender nothing is irrevocable, drawings can be redone, the scheme can be reduced, the project can be abandoned or restarted. At this stage any costs incurred for abortive feasibility and design work are relatively small compared to total cost of project (however on major and mega projects these costs can be more significant).

Furthermore, there is often no commitment to tenderers or need to compensate them for their trouble. However, clients should not be tempted to abuse the system since contractors' tendering costs will need to be recovered and will generally be reflected in more expensive projects.

Once the contract has been agreed between the employer and the contractor and the tender sum established the design should be frozen. It will then be necessary for both the client's and the contractor's project management (PM) teams to implement an efficient cost control system for the project.

The client's system should reflect the current financial status as well as the total anticipated price to be paid to the contractor. The contractor's system should reflect the current status as well as forecasting the anticipated total costs compared to anticipated total price to be paid by the client.

In the first half of this chapter we examine a client's post-contract cost control system, and in the second half a contractor's cost control system.

Employer's post-contract control

Basis of the cost control system

The post-contract cost control system should reflect the final figure which the employer will actually have to pay and for which he will have to raise and service finance. The system should continue to the point of completion/handover of the project.

Once the contract commences any future options are proportionately reduced – it being far more expensive to initiate major changes without substantial additional costs being incurred as a result of the delay and disruption caused.

The format of a post-contract control system can differ, depending on the type of contractual arrangements. However, the fundamental principles of any successful system remains the same:

- The system should allow amendment of the tender sum only after authorisation by the client's project executive.
- There should be a standard reporting system and all participants should be familiar with the procedures.
- All changes which affect the tender sum (design, construction, claims and commercial – above say £5000) should be agreed by the client's project executive before the instruction is given to the contractor.
- The client's PM team should rigorously implement such a procedure from day one; this will require a formal submission to the client's project executive covering all changes and instructions; the submission should be made in a standard format.
- Overall review of project reported at regular intervals to the client's project executive (monthly).
- The client's project executive should, at all times, be kept informed of the realistic anticipated final cost of the project (there should be no nasty surprises at the end of the project!).

Aims of a cost control system

The aims of the client's cost control system are:

- to enable the client to approve changes to the tender budget before such instructions modifying the works are given to the contractor;
- to enable the client to budget effectively for the anticipated expenditure;
- to enable the cost effect of any major change to be seen in the context of the project as a whole;
- to enable avoiding action to be taken if the total cost appears to be escalating unduly.

Typical client's contract cost control system

The system as described was designed and used by the HKMTRC and had a major impact on the successful outcome of the project, i.e. completion on time and within budget. Under the system no commitment would be entered into by the client without approval using the standard procedures.

Authorisation for all work likely to result in amendment to the 'contract control total' (CCT) could be made only with the prior approval of the client's project executive. The CCT was basically the tender sum less contingencies with prime cost/provisional sums and dayworks separately identified. The approval for any amendment was secured on a regular basis (client's project executive met twice a week) by the use of a Form 'X' application (see Figs 10.1 to 10.3).

West Midlands Major Project Corporation
Contract Cost Control

Proposal for a Form 'X' Submission

                                                    * Design
                                                       Construction
                                                       Claims
                                                       Commercial

Date: _____

To: _____

      *Design Manager/Contr. Admin./Senior Contracts Engineer

Copy:

Contract No. _____

I recommend that a Form X be raised in respect of the attached.

Proposer: _____   Title/Location: _____

To: _____   Date: _____

     * Chief Engin/Construction Manager/Contracts Manager

Copy:

My *Comments/Draft Form X in respect of the above are attached.

+ Input obtained from:

Signed: _____

    * Des Man/Con Admin/Sen Con Eng

To: _____   Date: _____

    * Engin Man/Proj Man/Comm Man
    Form X No. _____

Signed: _____

    Chief Engin/Construction Manager/Contract Manager

* Delete as necessary
+ Enter initials of Corporation Departments/Consultants consulted

Figure 10.1   Client's control document – Form 'X', page 1

West Midlands Major Project Corporation **Contract Cost Control** **Contract No. 829**	Form X Construction X/Cn/829/054 24/10/93

Proposal:

Dudley Town Station passenger adits
– provision of lead caulking

Cost range:

+ £27 500.00

Raised by:	Endorsed by:		Finance Div.	Principle Accepted
Const Man	ProjMan	ProjDir	FinDir	Exc.Comt.

Engineer's Instruction

E.I. No. _____ Issued

_____
Constr. Man.

Date: _____

Variation Order
By: Construction Manager

Indicate any significant programme or
other contractual effects arising since
above justification

Draft VO attached
Cost £_____

Construction Manager _____

Date _____

Commercial Aspects checked _____ Con.Man.

Recommended by:		Finance Division	Approved
PM	PD	FD	Exc Comt

Variation Order No. _____

_____
Construction Manager

Issued: _____

Figure 10.2   Client's control document – Form 'X', page 2

Contract 829
Form X/CN/829/054

Dudley Town Station
Passenger adits – additional lead caulking

Justification

The adits in Contract No. 829 are constructed with cast iron segmental linings.

The design of the segments does not allow for a sealing strip and watertightness is dependent on the perfect contact of the segment faces and grouting.

The B of Q item for the segmental rings included for sealing strips – this being at variance with the finished design.

It is anticipated that the adits will move as the adjacent excavation in Contract No. 830 (Dudley Town Concourse) progresses. This will cause leakage due to the opening of the joints and damage to the grout – a long-term leakage problem will therefore be likely.

It is therefore recommended that the adits be fully caulked with lead caulking. This will make for a more effective seal should movement of the adits occur and will also make for easier longer-term maintenance.

Provisional items were included in the contract for caulking to these linings.

The total cost of the caulking, based on the rates included in the B of Q, will be £27 500.00 (as appendix 'A' enclosed).

This Form X was not included in the September 1993 Unit Financial Report.

KFP/JD/ep
24.10.93

Figure 10.3   Client's control document – Form 'X' (Justification)

Following the contract award applications to raise a Form 'X' should always be made in connection with:

- prime cost or contingency sums;
- instruction of additional work (either temporary or permanent);
- issue of working drawings or information likely to modify the client's liability due to admeasurement;
- contractor's claims submissions;
- modification to key dates/milestones or liquidated damages;
- acceleration of the works.

Engineer's instructions should first be issued instructing the contractor to execute the work prior to the issue of an official variation order, the latter of which should reflect the full cost/time implications.

Form 'X' submissions, which are submitted to the client's project executive before instructions are given to the contractor, normally arise from one of three sources:

1. *Design Form 'X'*: Covers all proposals to change or modify the design or specification within the contract documentation and the placing of nominated subcontractors/suppliers.

   The Form 'X' is raised by the designer, then approved by the appropriate line managers – design manager, chief engineer and the engineering manager; however, the authority to commence the work must await the endorsement by the project director on p. 2 of the Form 'X' (Fig. 10.2).

2. *Construction Form 'X'*: Covers all instructions covering additional temporary work and changes in the sequence/method of construction of permanent works which result in a change to the contract price.

    The Form 'X' is raised by the site quantity surveyor, then approved by the appropriate line managers – contract administrator, construction manager and engineer/project manager; again the authority to commence the work must await endorsement by the project director.

    However, construction managers, who have been delegated extensive powers under the contract, may issue instructions up to the value of say £10 000 each month without recourse to the above procedure.

3. *Claims/Commercial Form 'X'*: Covers all claims notified by the contractor under the contract and any extra-contractual arrangements, for example acceleration measures or alteration to key milestone dates. The Form 'X', with a full report on the claim identifying the client's potential liability, should be raised immediately the contractor issues an initial notice to claim.

    The Form 'X' is raised by the contracts administrator/senior contracts engineer with assistance from site quantity surveyor, then approved by appropriate line managers – construction manager/contracts manager and commercial manager; again the authority to respond to the contractor and make payment must await the endorsement of the project director.

At the end of each month, after finalising the interim valuation, the client's quantity surveyor should prepare the monthly financial statement which monitors the control of costs under the various cost heads. This statement should be included in the monthly report prepared for each contract and submitted to client's project executive (Fig. 10.4).

Contractor's post-contract control

Introduction

The second part of this chapter is based on an actual review of a construction company's cost control procedures. The purpose of the review is to improve the efficiency and effectiveness of the company's cost control procedures by looking at the current management theory and developing the existing system.

The review of the Mightybuild Construction Company cost control system was requested by the construction director as part of the ongoing review and updating of the company procedures as set out in the company's Five-Year Plan.

Market conditions

The construction industry in the UK is currently in a period of recession. The recession emphasises the need for efficient and effective cost control

West Midlands Major Project Corporation   Contract No. 829   Financial statement No. 034 as at September 1993

Cost centre	Cost centre title/ activity bill	Control total	Potential requirements		Form X approvals (final)				Expected final contract cost (3±4 to 9)	Net (Savings) Excess 10−(3 + 4)
			Provisional sums and items	Variations, remeasure and PS/PI increases	Provisional items	Provisional sums	Variations and dayworks	Remeasure		
1	2	3	4	5	6	7	8	9	10	11
A	Preliminaries	11 200 000	—	150 000			40 000		11 390 000	190 000
	–provisional sums	—	460 000	—					460 000	—
B	Dudley station	18 900 000	—	(170 000)			(50 000)		18 680 000	(220 000)
	–provisional items	—	500 000	(160 000)	(80 000)				260 000	(240 000)
C	Dudley–Tipton running tunnels	12 500 000	—	72 000			85 000		12 657 000	157 000
	–provisional items	—	460 000	(100 000)					360 000	(100 000)
D	Tipton–Dudley running tunnels	18 000 000	—	(70 000)			800 000		18 730 000	730 000
	–provisional items	—	700 000	(550 000)					150 000	(550 000)
E	Tipton station	19 500 000	—	300 000			20 000		19 820 000	320 000
	–provisional items	—	25 000	—					25 000	—
	Initial contract total	80 100 000	2 145 000	(528 000)	(80 000)		895 000		82 532 000	287 000

Claims settled (not included above) Commercial agreements   1 600 000 / Nil   1 600 000 / Nil

Expected final contract total   84 132 000   1 887 000

	Control total
Provisional sums	460 000
Provisional items	1 685 000
Dayworks	750 000
Prime cost items	—
Contract sum	82 995 000

Contractors' stated value of other notified claims	4 000 000
Claims expected	200 000
Engineer's estimate of settlement of other notified claims	1 500 000

Figure 10.4   Client's control document – monthly financial statement

to enable the company to survive on greatly reduced tender margins until conditions improve. When conditions do improve an efficient and effective cost control system will enable the company to maximise profits.

Management theory

Traditionally cost control in the construction industry has been taken from manufacturing and service industries. It has taken recorded costs and compared them with the budget. Thus it has been historical in nature and of little or no use in controlling costs even where this was possible because either the information was out of date before it was available or the operation to which the information referred had been completed.

Current management theory considers that effective cost control can be realised only if it is forward-looking rather than backward-looking. Therefore cost forecasting is needed rather than cost recording. This change of emphasis allows management to make choices regarding resources or methods of working before committing the company, having anticipated the financial consequences and incorporated the costs in the construction budget.

The above approach to cost control has a secondary benefit in that it conditions management to look forward in time, this being an area where there is at least some chance of exercising control, instead of looking to the past where there is no chance of doing so.

Fundamental rule of project management:
'You cannot control the past – only the future.'

Existing system – statement

The existing company cost control system is accounting based and historical in operation. The contract costs are allocated on each project in terms of materials, domestic subcontractors, nominated subcontractors, labour-only subcontractors, direct labour, plant and transport and preliminaries. These cost centres are further subdivided in the master cost control system and utilised as required on individual projects. The number of subdivisions varies from 9 for direct labour to 35 for domestic subcontractors.

The contract costs are published monthly for each project with cumulative and monthly costs for the whole project, cost centres and cost divisions. The published costs are adjusted manually to bring all the costs to the same date as the external interim valuations agreed with the employer's quantity surveyor or engineer.

The manual adjustment of costs involves deducting or adding the cost of materials, labour, construction equipment and preliminaries included in the published costs. Deductions are made for those resources which occurred or were used after the nominated valuation date and additions made for those resources which were used before the valuation date and for which no costs have been included in the published costs. For domestic and nominated subcontractors the manual adjustment usually comprises the addition of costs for liabilities not included in the published cost.

The value of work completed and materials on site at the nominated valuation date is then calculated to give the turnover figure for the project. If the external valuation is used as the starting-point this is manually adjusted for over- and undervaluation of measured work, variations and materials on site as well as any invoices which have been raised for work outside the contract.

The adjusted costs are compared with the turnover figure to determine the actual profitability of the project. Any significant variance between the actual and budgeted project profitability should be investigated to determine its main cause or causes and any appropriate and possible corrective action taken and feedback to site management, estimating, planning and buying departments initiated.

System problems

The company cost control system is accounting based and historical in operation. Both of these features reduce the effectiveness of the system for the following reasons:

- Because the system has been developed by, and is peripheral to, the accounts department it is perceived by certain contract personnel as less important and less relevant than other company systems such as planning and subcontract administration.
- Because the system is accounting based the costing detail produced is of limited use unless a lot of time is spent analysing the data into different cost arrangements, such as excavation, concrete costs, etc.
- The level of some cost detail is too great. This does not facilitate its synthesis into the different cost arrangements noted above and is of no practical use to the contracts department.
- Because of the large number of headings available and the allocation of certain ambiguous costs by the accounting clerical staff costing errors can be easily made.
- The manual adjustment of the costs at the end of the month mitigates against the reorganisation of the data into the different cost arrangements noted earlier.
- Because the costs are published monthly and 8 working days after the month end the information can be up to 6 weeks out of date.
- If an item was identified that was not meeting the required profit and corrective action was possible it would be likely that either the operation would have finished or would have changed to such an extent that any corrective action initiated by the cost control system would not have the desired effect.
- The arbitrary allocation of head office contract costs to individual projects can lead to the situation where costs are allocated to those projects that can afford them rather on the basis of actual time and cost.
- The manual adjustment of the value of work carried out generally leaves the information in the same format as the originating document, usually the bill of quantities (B of Q). The resulting data can be

compared with the cost data only as an overall figure, unless already either in, or rearranged into, the required cost arrangement as noted earlier.

- The inclusion of the value of materials on site in the turnover figure for a project reduces the profitability of a project because the materials are valued at net cost. To correct this problem, which may be substantial where a project has a high proportion of its turnover as materials on site, the value needs to be adjusted, usually by the average profitability of the project. This adjustment can itself be problematical in that profit would actually be taken before the work was carried out and the actual work is unlikely to be as profitable as the average profitability of the project.

- If the overall profitability of a project is significantly different from the budgeted profitability it is very difficult to determine the cause or causes without the lengthy and time-consuming task of manually analysing the costs and turnover into cost centres.

Proposed system

If a contract is to be completed within budget the financial effect of all decisions must be forecast before the decision is made. Accurate forecasts based upon correct information should be made using cost records either from the project itself or from others. Each decision will have three stages:

1. The options available have to be set out.
2. A cost forecast has to be made for each option.
3. Having made the decision, the total cost forecast has to be updated to include the chosen option, taking into account any unexpended contingencies.

Where it is forecast that the chosen option will exceed the budget the contingency will be released to compensate for the difference. Where it is forecast that the chosen option will be less than the budget the surplus money will be added to the contingencies.

The initial magnitude of the contingencies on any project will depend upon factors such as the degree of contractual financial risk, the expertise of the contract team, the level of inflation, the nature of the project, etc.

The contingencies are to be used on the instruction of the project manager to supplement the budget of those operations which have been identified as either potential or actual loss makers. The adjusted contingency sum allows the project manager to determine at any point during the contract the financial position of the contract and the amount of financial flexibility available for any future decisions that may be required.

Every option considered will have time and performance implications and these factors together with any knock-on effect on subsequent operations need to be taken into account when making any decision.

The project cost records may be used to verify the accuracy of the individual option forecasts or the overall cumulative total cost forecast for future forecasting and to adjust the total cost forecast as necessary.

Key areas for consideration

To enable the proposed system to assist management to keep within budget the following key areas need to be addressed:

- Managers and their systems must give prompts when a decision is needed.
- All possible options for any decision should be considered.
- Good cost forecasts of all possible options must be made.
- The chosen option must be the 'best fit' into the total cost forecast.
- The total cost forecast must be regularly updated so as to be ready for use when the decision is prompted.

Completing within budget depends upon the combination of three factors:

1. An ability to make prompt control decisions;
2. An ability to make accurate forecasts;
3. An effective policy for setting and controlling contingencies.

The above factors will require further management input to set up the necessary systems and determine the appropriate policy.

The cost recording system should have the following features:

- simplicity of operation;
- be integrated with the construction programme;
- exhibit integrated value and cost;
- budget and actual figures should be differentiated for comparison purposes.

The basis of the proposed cost recording system is either the analytically priced B of Q for traditional contracts or the contract sum analysis for design and build contracts together with the contract programme.

The total cost forecast should be distributed across the programmed operations to allow actual turnover to be compared to the planned, or budgeted, turnover. This should be done at monthly intervals based upon actual progress as compared to the planned, or budgeted, progress.

It is extremely important that the estimate should be realistically priced, using appropriate outputs and rates derived from actual past performance and work study procedures where these are available.

The expertise of the contracts and planning departments should also be utilised to ensure as far as possible that the most cost- and time-effective methods are allowed and priced for within the estimate.

The tender programme should be resourced and costed as a check on the labour and construction equipment content of the estimate and any discrepancy between the estimate and the costed tender programme should be corrected before the tender adjudication meeting.

Short-term planning and control

Professor Geoffrey Trimble, at Loughborough University of Technology, devised a simple forward-looking short-term planning and control scheme in conjunction with Bovis Civil Engineering.[1] Under the scheme the contractor's section engineers prepared a short-term plan showing the proposed activities for a section of the works together with the cost and value for that section. The cost and value were closely monitored on a weekly basis and could be combined with the calculation of payments for an operatives' bonus scheme or for subcontract labour.

Under the scheme cost is largely controlled in advance by simulating the possible alternative plans and quantifying their cost implications ('what-if' calculations can more easily be made using a computer). If the costs are excessive for the work when executed the section engineer should implement any corrective action immediately.

The scheme has several plus factors: it is forward-looking and enables corrective action to be taken immediately, it ensures that adequate records are kept for each section of the work – particularly critical if the work becomes subject to a contractual claim – and section engineers become aware of the cost of operations – particularly relevant where large items of construction equipment or standing scaffold are involved.

On building projects the value calculation may, in practice, be more time-consuming to calculate thus making this forward-looking short-term planning and control scheme less economic to operate.

Establishing the cost centres

The number of cost centres on any project will be determined by the degree of information required by management and the resources required to operate the system. The cost control system must be cost beneficial. Therefore as a general rule, the number of cost centres should be kept to the minimum necessary to give adequate and appropriate control.

As the B of Q or contract sum analysis on any project is unlikely to be in a format that can be used for cost control purposes it will usually be necessary to analyse the information and to rearrange it into the required cost control format. This will entail making arbitrary decisions as to the boundaries between cost centres and allocating the bill rates accordingly. It is important that these boundaries are well defined and communicated to the necessary personnel so that the subsequent allocation of resources and therefore costs is correctly carried out, thus avoiding, as far as possible, unproductive corrective action.

The incorporation of variations into the cost control system may prove to be a problem because of the need to integrate the differentiated value and cost elements of the individual variations with the original information on a regular basis. This will probably necessitate initially inputting the data on a provisional basis and later updating the data when the variation has been agreed.

Attributes of an effective system

An effective system of construction cost control should exhibit the following characteristics:

• A budget for the project should be set with a contingency figure to be used at the discretion of the responsible manager.

- Costs should be forecast before decisions are made to allow for the consideration of all possible alternative courses of action.
- The cost recording system should be cost-effective to operate.
- Actual costs should be compared with forecasted costs at appropriate periods to ensure conformity with the budget and to allow for corrective action if necessary and if possible.
- Actual costs should be subject to variance analysis to determine reasons for any deviation from the budget for use by estimating or bonus departments.
- The cost implications of time and quality should be incorporated into the decision-making process.

Cost recording models

The type and sophistication of any cost reporting system will be determined by the resources available to operate the system and the use made of the system by the relevant management personnel.

Harris and McCaffer[2] suggest that a coarse-grained system using no more than fifteen cost codes be used. Unpublished research supports the view that when large numbers of codes are used recording cost data becomes highly erroneous. They also demonstrated that there is a break-even point on any project at which the cost of any extra control effort will be uneconomic and increase the overall project cost.

Briscoe[3] describes a simplified form of a standard costing system that allocates costs for different activities or elements of the total project. Sufficient information is available to permit costs to be subdivided into direct materials, labour, plant and subcontractors with site overheads distributed across each element. As the contract proceeds the actual costs are compared to the budgeted values and the variances identified.

Pilcher[4] discusses the use of an integrated reporting system to control the scheduling and cost control processes. He concludes that separate schedule and cost control systems provide a cheaper means of control and the output from the two separate systems is more easily understood.

Figure 10.5 shows a typical contractor's monthly cost report for a jetties contract on an international marine terminal. The report is prepared on a monthly basis following the interim valuation agreed with the client. The report requires a comparison to be made between the value of the work done and the cost of doing it, i.e. the variance. The aim of the report is to identify problem areas and trends as well as forecasting the final profit/loss situation on the project. Corrective action should be taken on any cost centres showing a loss if at all possible. This form of cost report is most effective on a contract with repetitive operations but is less so on non-repetitive contracts.

The time for completion of the project in Fig. 10.5 is 90 weeks with a contract value of £11.9m. The cost report reflects the financial situation at the end of week 60 and it is noted that the project is 5 weeks behind the tender programme, the revised finishing date being week 95.

The general items, or preliminaries have been split into two items representing:

Mighty build Construction Company
Contract: Begawan jetties
Contract No. 92/6

Monthly cost report
Date: end Jan 1994
Prepared by: K. F. P.

Programme : as tender : 90 weeks
To date : 60 weeks
Actual : 55 weeks

Cost code	Description of Work	Unit	Quantity				Cost (£1000s)						
			B of Q	To Date	Estimated Final Total	To Complete	B of Q	Cost to Date	Valuation to Date	+ To – Date	Estimated Final Cost	Final Valuation	Final + –
010	General items	Item	Fixed	90%	100%	10%	1 420	1 380	1 278	(102)	1 535	1 420	(115)
		Weeks	Time related	60	95	35	1 080	810	660	(150)	1 283	1 080	(203)
020	Piling	No.	750	550	780	230	1 875	1 250	1 375	125	1 773	1 950	177
030	Structural steelwork	tonne	1 500	900	1 570	670	4 500	2 250	2 700	450	3 925	4 710	785
040	Precast concrete	m3	1 500	750	1 530	780	750	490	375	(115)	1 000	765	(235)
050	*In situ* concrete	m3	700	300	750	450	175	120	75	(45)	300	188	(112)
060	Pipework	m	1 800	500	1 950	1 450	1 350	295	375	80	1 150	1 463	312
070	Loading arms	No.	6	Nil	6	6	450	—	—	—			—
080	Electrical	Item	Fixed	Nil			150	—	—	—			—
090	Buildings	Item	Fixed	Nil			150	—	—	—			—
							11 900	6 595	6 838	243	10 966	11 576	610

Figure 10.5   Contractor's control document – monthly cost report

1. Fixed charges – cost of mobilisation/demobilisation of marine construction equipment together with the contractor's land-based facilities.
2. Time-related charges – representing the weekly cost of running the site including costs of all staff, maintaining the construction equipment and fuel.

Of the 'fixed charges' within the general items 90 per cent has been paid to date (£1 278 000) and this is considered a realistic assessment of the situation; however, the actual costs are shown as (£1 380 000), thus showing a loss to date of (£102 000).

All the estimated final costs for each cost code have been calculated pro rata to the actual costs to date. The final valuation figure is based on the initial tender figure adjusted if necessary for any variations, i.e. both the final cost and valuation are calculated representing the final estimated quantity.

The client's contract administrator has paid for only 55 weeks of 'time-related' costs – reflecting the actual progress to date. The report shows a final valuation of £1 080 000 (as tender), with the final costs based on 95 weeks, i.e. £1 283 000 – an anticipated final deficit of £203 000.

It could be that the 5-week delay was due to the contractor encountering unforeseen ground conditions; then some, or all, of the deficit may be recovered through the submission of a contractual claim made under the appropriate clauses in the conditions of contract (e.g. FIDIC clauses 12 and 44).

Furthermore, the losses on the precast concrete work could be due to one or more of several reasons: the B of Q rates were undervalued, breakages and waste have been excessive, the basic method used in the precasting yard is uneconomic, e.g. involving too much handling or not enough use of standard shutters perhaps indicating a design change, or the work has been delayed due to late instructions from the engineer. On this particular project it is known that the engineer has issued information on the precast units late and in a piecemeal fashion which has caused severe delay and disruption in the contractor's precasting yard.

If the contractor can prove that his original estimates were realistic and that he was achieving profitable production rates prior to the problems with the late issue of drawings then the monthly cost report could prove to be a most useful document in the substantiation of any claims.

Acknowledgement

Much of the material for the second half of this chapter was based on an actual critical review of a construction company's cost control procedures undertaken by Robert Church under a 'project management' assignment on the M.Sc. Construction Management course at the University of Wolverhampton (award made jointly with Loughborough University).

Tutorial questions

1. You have been asked by your contracts director/senior partner to review critically one of your organisation's areas of control. Prepare a report (not exceeding 3000 words) on one of the following:
    (a) time control
    (b) cost control
    (c) quality control
    ('Project management' assignment, 1991, M.Sc. Construction Management, University of Wolverhampton)
2. Work obtained by the selective tendering process demands a high level of cost control by production management during the construction period. Discuss methods and techniques which can be used in the pre-tender and construction periods as an aid to financial control. (CIOB Member Examination Paper II, Contract Administration, 1988)
3. Explain the extent to which post-contract cost control is dependent on, or influenced by, pre-contract tender procedures. (RICS Final (1987 syllabus), specimen paper)
4. Discuss the usefulness of 'S' curves in post-contract cost control. (RICS Direct Membership Examination, Project Cost Management, 1989)
5. This assignment relates to a financial control system and answers should be supported by proper pro forma documentation including a typical worked example to illustrate its effectiveness as a control measure.

    The student group should begin by identifying the control information desired from such a system; devise a flow chart or similar to illustrate its components and their source. Subsequently full design of the system can take place.

    Financial control must be treated in its widest sense to include not only aspects of budgeting and costing but also the establishment of criteria to assess managerial performance and corporate status.

    A single submission, based on one assignment only, will be required from each group but it must contain four sections:
    *   statement of control criteria
    *   system design
    *   pro forma documentation
    *   an example of application

    The group should work as a team but each member will be nominated to assume leading responsibility for a contribution in one of the four areas. The submission should clearly identify these definitive roles.

    Upon completion the group will be required to present their results by a 10-minute viva for peer scrutiny and challenge.

Assignment 1
Produce such a system as the financial basis of management control for an individual construction project (contractor's system).

Assignment 2
A system for executive management for an individual construction project (employer's system).

Assignment 3
A system for executive management of a construction firm.

Assignment 4
A system for the executive management of a construction professional practice (architect/quantity surveyor/civil engineer).

('Finance' assignment, Level 3, construction degree programme, 1993, University of Wolverhampton)

References

1. Trimble G, Neale R H, Backus S 1985 Effective control of project costs, within 'Project control selected reading', M.Sc. Construction Management course notes, Loughborough University of Technology.
2. Harris F, McCaffer R 1989 *Modern Construction Management* 3rd edn BSP Professional Books
3. Briscoe G 1988 The *Economics of the Construction Industry* Mitchell
4. Pilcher R 1985 *Project Cost Control* Collins

Further reading

Pilcher R 1992 *Principles of Construction Management* 3rd edn McGraw-Hill

Variations

Introduction

A most important post-contract duty for the client's contract administrator or the contractor's commercial manager is the evaluation and subsequent agreement of variations.

Variations are inevitable on major construction projects particularly where the project is complex, the ground conditions variable or the promoter has a desire to incorporate the latest technology into the scheme. On large projects the number of variations may run into thousands, often valued at millions of pounds. It is not surprising therefore to learn that the valuation of variations is one of the commonest sources of dispute in the construction industry.

Indeed on some projects it may be more economic in the long term for the promoter to 'freeze' the design at the commencement of the project with all changes being treated as separate contracts on completion, though of course this is not always practical.

Principles of valuation

The valuation of variations requires experience, sound judgement and integrity. It requires that the parties have a detailed knowledge of methods of construction, estimating practice and costs; often it will involve consideration of the effect of the variations on the programme of the works.

If the varied works are complex the parties need to be skilled negotiators and be prepared to adopt a give-and-take attitude in order to bring about a satisfactory settlement. Compromise is often required for there is seldom one correct solution. Indeed both parties may consider several different approaches before selecting the appropriate strategy.

The great majority of contract provisions for additional payments for variations employ the concept of price, not of cost, when providing for their valuation. The intention will normally be to use prices which, in a fixed price contract, will be related to the estimated level of prices at the likely time of performance of the work. In a contract with a fluctuation of price provision care should be taken to ensure that reimbursement for increased costs is taken only once.

Clause 52(1) of the ICE 6th requires that the engineer establishes the value of variations whilst clause 13.4.1.1 of JCT 80 requires the quantity surveyor to establish this value; both are required to consult with the contractor before the price is established. In practice it is often the contractor who first establishes the new rate, based on the records and costs available to him, before submitting it to the engineer/quantity surveyor for agreement.

The principles for the valuation of variations in clause 52 of the ICE Conditions of Contract (6th edition) and clause 13 of the JCT Standard Form of Building Contract (1980 edition) are not dissimilar (see Fig. 11.1).

Bill of quantities rates are used to value variations where the varied work is of a similar character and executed under similar conditions. Where the work is not executed under similar conditions the prices in the B of Q are used as a basis for valuation 'so far as may be reasonable', failing which a 'fair valuation is made'.

When considering whether a variation is executed under 'similar conditions' it is necessary to consider all the circumstances which may have a bearing on the cost – all those factors which should reasonably have been included at tender should be compared to the actual conditions.

ICE 6th	JCT 80
Ascertained by the engineer after consultation with the contractor (52(1))	Ascertained by the quantity surveyor after giving the contractor the opportunity of being present (13.6)
Where work is of similar character and executed under similar conditions to work priced in the B of Q it shall be valued at such rates and prices contained therein as may be applicable (52(1)(a))	Where the additional or substituted work is of similar character to, is executed under similar conditions as, and does not significantly change the quantity of, work set out in the contract bills the rates and prices for the work so set out shall determine the valuation (13.5.1.1)
Where work is not of a similar character or is not executed under similar conditions or is ordered during the defects correction period the rates and prices in the B of Q shall be used as the basis for valuation so far as may be reasonable, failing which a fair valuation shall be made (52(2)(b))	Where the additional or substituted work is of similar character to work set out in the contract bills but is not executed under similar conditions thereto and/or significantly changes the quantity thereof, the rates and prices for the work so set out shall be the basis for determining the valuation, and the valuation shall include a fair allowance for such difference in conditions and/or quantity (13.5.1.2)
	Where the additional or substituted work is not of similar character to work set out in the contract bills the work shall be valued at fair rates and prices (13.5.3)

Figure 11.1   Principles of valuing variations. *Sources:* ICE Conditions of Contract 6th Edition, reproduced by permission of the Institution of Civil Engineers, and JCT 80 Private with Quantities Contract, reproduced by permission of the copyright holders © RIBA Publications Ltd (1980)

Dayworks

Under the JCT 80 if the work cannot be measured, or under the ICE 6th if the engineer considers 'it is necessary or desirable', the architect/engineer should instruct the contractor to keep contemporary records and the work should be valued on a daywork basis.

JCT 80 clause 13.5.4 requires that the prime cost of dayworks is calculated using the *Definition of Prime Cost of Dayworks Carried out under a Building Contract* issued by the RICS and the Building Employers' Confederation current at the base date.[1] Rates for construction equipment should be based on the RICS *Schedule of Basic Plant Charges*.[2]

Under the JCT contract where the work is within the province of a specialist contractor, e.g. electrical or heating and ventilating, those definitions agreed between the RICS and the appropriate body representing the employers should be used.

Clauses 52(3) and 56(4) of the ICE 6th require that dayworks shall be reimbursed under the conditions and at the rates and prices set out in the daywork schedule in the contract. Failing the inclusion of such a schedule in the contract the contractor will be reimbursed at the rates and prices under the *Schedules of Dayworks Carried out Incidental to Contract Work* issued by the Federation of Civil Engineering Contractors (FCEC) current at the date of execution of the daywork.[3]

As the FCEC Schedule contains a far more comprehensive list of construction equipment than the RICS list it may be a useful guide even on building contracts under JCT 80. *The Surveyors Guide to Civil Engineering Plant* published by the Institution of Civil Engineering Surveyors[4] provides a useful checklist for identifying items of construction equipment against the FCEC plant reference number.

It is noted that the FCEC Schedule is designed for use when valuing dayworks incidental to the contract and may not be appropriate for valuing major variations or for those instructed after substantial completion.

Daywork records are often worth keeping and agreeing on a 'without prejudice' basis to facilitate any valuation that may later become necessary.

Other relevant clauses

JCT 80 clause 13.5.5 and ICE 6th clause 52(2) may also be relevant in connection with variations. Under these clauses the parties may review the rates of any other work which is executed under different conditions from those anticipated at tender following the ordering of a variation.

Furthermore, ICE 6th clause 56(2) requires no order in writing to be issued for an increase or decrease in quantities when the remeasured quantities are different from those described in the B of Q unless due to a variation. Clause 56(2) thus relates to the correction of errors in quantities in the B of Q and not to changes due to a revised design indicated on drawings.

Under the ICE 6th clause 56(2) if the engineer considers that the remeasured quantities have rendered any of the rates inappropriate then they may be adjusted up or down after consultation with the contractor. This may be particularly relevant where items for mobilisation/ demobilisation of construction equipment or temporary works have been included in the rates or where a reduced quantity makes the use of large construction equipment uneconomic.

The engineer under the ICE 6th or the architect under JCT 80 should order variations to the works only through the powers vested in them under the conditions of contract. If changes are required to the contract itself this should be a matter for negotiation between the employer and the contractor, e.g. acceleration under JCT 80. Negotiation between the parties may also be necessary where the changes are so substantial as to alter the basis of the original contract.

Form of instruction

All variations should be issued in writing normally under the cover of an official variation order. It is not considered good practice for the engineer/architect merely to issue drawings which conceal the fact that there are variations. The employer will need to know the likely cost of these variations and give approval before the instruction is given and the contractor will obviously require payment.

Any verbal instructions received from the engineer/architect should be confirmed in writing as soon as possible. In practice it is often the contractor who promptly confirms the oral instructions, for if these confirmations are not contradicted within a very short period of time they become valid instructions.

Basis of rate fixing

At the commencement of the project the parties should agree a rational and consistent policy for the procedures and methods to be adopted in fixing new rates.

Charles Haswell in his paper entitled 'Rate fixing in civil engineering contracts' published in the ICE *Proceedings* in 1963[5] considered that the rate for a variation should be the rate that the contractor would have inserted against that item had it been included at the time of tender. This approach would seem correct; however, the problem for the contract administrator is in establishing the breakdown of the rates in the B of Q so that new rates may be built up in a similar manner. Each rate normally consists of three parts:

1. The prime cost including:
   (a) labour: usually based on all-in hourly rates, outputs from standard price books should be used with caution as they may not be appropriate to the particular circumstances;
   (b) material (permanent): if taken out of stock include handling charge, waste percentage often far more in actual practice than included in tender;
   (c) material (temporary): designed by contractor and often of substantial value, number of uses on formwork may change with variation;

(d) construction equipment including operatives: estimated based on continuous use. Fuel and transport on and off site may be included in the rate or in preliminaries;

(e) subcontract costs: based on subcontractor's quotations – main contractor's discount and special attendances may or may not be reflected in the rates.

2. The site overheads including time-based elements such as staffing, welfare facilities and fixed elements such as insurances and mobilisation of facilities.

3. Head office overheads, risk and profit.

The difference between item 1 – the prime cost – and the final bid figure is called 'the spread' and it is this amount of money which has to be distributed amongst the rates.

There are number of ways in which the spread can be distributed and a different approach can often explain why the same item in different contractors' bids shows a wide range of prices. The spread could be distributed evenly on all the rates, it could be spread unevenly, e.g. front-end loaded, or it could be included as a lump sum in the preliminaries or general items.

A further point worthy of note is the difference in approach between building and civil engineering estimating. In the former the rates in the B of Q are usually established by the unit rate method, i.e. based on 1 m² of permanent works. In civil engineering estimating the operational method is normally used in which the total cost of the operation is calculated and then split down and inserted against the appropriate descriptions and quantities in the B of Q as described in Chapter 8.

Fixing the rate

The rates in the B of Q should be the basis for the valuation of variations so far as may be reasonable. Max Abrahamson reflects in his classic book *Engineering Law and the ICE Contracts*[6] that it is not reasonable to argue that such rates should not be used because they are mistaken or uneconomic. He further considers that if the contractor has struck a bad bargain it is not the intention to permit him to make up such loss through the excessive valuation of variations. Only in exceptional cases should the rates in a B of Q be abandoned.

However, the contract administrator's dilemma is in identifying the breakdown of the bill rates in order to make a pro rata valuation as the contractor would not normally be required to submit a breakdown of his bill rates with his tender. In contrast it is understood that the VOB contracts in Germany and the SIA contracts in the Singapore private sector require contractors to submit a detailed breakdown of their rates.

The client's contract administrator can overcome this problem to some extent by asking the successful contractor at the commencement of the project for a detailed breakdown of the significant rates. A breakdown of say six of the main items covering excavation, concrete, formwork and reinforcement may be enough to establish the contractor's method of

Item in bill of quantities:

F133   Provision of concrete standard mix, ST3, cement to BS 12 20 mm aggregate

£56.72/m³

N.B. Placing of concrete measured as a separate item as CESMM3 rules.

Engineer's Instruction Nr. 23 deletes provision of concrete mix ST3 and replaces with Mix ST5

Calculate rate for new item:

F153   Provision of concrete standard mix, ST5, cement to BS 12 20 mm aggregate

? / m³

Data on concrete mixes as BS 5328 : Part 2 : 1990 Section 4

Table 5. Mix proportions for standard mixes					
Standard mix	Constituent	Nominal maximum size of aggregate			
		40mm		20mm	
		slump 75mm	slump 125mm	slump 75mm	slump 125mm
ST1	Cement (kg)	180	200	210	230
	Total aggregate (kg)	2010	1950	1940	1880
ST2	Cement (kg)	210	230	240	260
	Total aggregate (kg)	1980	920	1920	1860
ST3	Cement (kg)	240	260	270	300
	Total aggregate (kg)	1950	1900	1800	1820
ST4	Cement (kg)	280	300	300	330
	Total aggregate	1920	1860	1860	1800
ST5	Cement (kg)	320	340	340	370
	Total aggregate (kg)	1820	1860	1830	1770
ST1 ST2	Fine aggregate (percentage by mass of total aggregate)	30 to 45	30 to 45	35 to 50	35 to 50
ST4 ST5	Fine aggregate (percentage by mass of total aggregate)				
	Grading limits C	30 to 40		35 to 45	
	Grading limits M	25 to 35		30 to 40	
	Grading limits F	25 to 30		25 to 35	

Specification indicates:

• nominal maximum size of aggregate = 20 mm
• 75 mm slump
• 40% fine aggregate

Figure 11.2   Example No. 1 – using B of Q rate to establish new rate

Analysis of materials element in 1m³ of concrete:

BS5328:Part 2:1990 indicates:

	ST3
Cement (kg) 270 @ £64/t	17.28
Total aggregate 1890 kg	
Fine aggregate (sand) (kg)	
40% × 1890 = 756 kg @ £8/t	6.05
Coarse aggregate (gravel) (kg)	
60% × 1890 = 1134 kg @ £10/t	11.34
	£34.67

Analysis of B of Q rate (Mix ST3)

	Labour	5.00
	Materials	34.67
	Plant	6.00
		45.67
Site overheads @ 15%		6.85
		52.52
Head office overheads and profit @ 8%		4.20
B of Q rate		£ 56.72 / m³

Build-up of new rate (Mix ST5)

Engineer specifies same aggregate size, slump and percentage fine aggregate as B of Q item.

Rate build up using same prices and mark-up as B of Q rate based on data taken from BS5328:Part 2:1990

Build-up of materials element in 1m³ concrete:

Cement (kg) 340 @ £64/t	21.76
Total aggregate 1830 kg	
Fine aggregate (sand) (kg)	
40% × 1830 = 732 kg @ £8/t	5.86
Coarse aggregate (gravel) (kg)	
60% × 1830 = 1098 kg @ £10/t	10.98
	£38.60

	Labour	5.00
	Materials	38.60
	Plant	6.00
		49.60
Site overheads @ 15%		7.44
		57.04
Head office overheads and profit @ 8%		4.56
New rate		£61.60 / m³

Figure 11.2　*Continued*

pricing, labour and plant costs and percentage for 'spread'. Unfortunately the contractor is not contractually obliged to comply with such a request, but it is worth stating that in the event of an arbitration the contractor may be required to submit all his original estimate sheets to the arbitrator.

Once the contractor's 'spread' is established the rate for the varied work can be calculated on the same basis as the original rate in the B of Q. Obviously if the rate is calculated after the work is done and is based on actual records and costs there would be no need to add a contingency or risk item.

Fair valuation

Where the varied work is completely different in character from that described in the B of Q it should be valued on the basis of a 'fair valuation' (ICE 6th) or 'fair rates and prices' (JCT 80). Unfortunately few commentators have given any opinion as what to what can be interpreted from these words.

Max Abrahamson considers that a 'fair valuation will normally mean cost plus a reasonable percentage for profit'; unfortunately he does not define what is meant by reasonable. He further states that 'if there is a proof of a general market rate for comparable work it may be taken into consideration or applied completely'.

The author recollects on one major project how the whole profitability of the project was influenced by the use of 'the general market rate' not for work done but for construction equipment. The problem revolved around establishing hourly rates for the contractor's own large mobile cranes (150 tonnes lifting capacity) when working on varied work. The contractor argued that his own internal rates were confidential and inappropriate. A theoretical rate based on the extrapolation of the FCEC rates (the crane capacity was larger than any in the FCEC Schedule) was also considered inappropriate.

Eventually the engineer and the employer accepted that the hourly rate could be based on the 'general market rate'; similar capacity hired cranes were in use on an adjacent site. However, the client may have regretted it later when the contractor used these market rates in a major claim for acceleration which involved the use of numerous large mobile cranes over an extended period. This approach may seem ideal from a contractor's viewpoint in times of economic boom when there is a high demand for major items of construction equipment; however it may not be such a good idea in times of recession!

An alternative method of establishing an hourly rate for those major items of capital equipment not included in the FCEC Daywork Schedules or the CPA Schedules, e.g. tunnelling shields, dredgers, crushing plant and large cranes, would be to build up the rate based on first principles. Consideration would need to be given to owning costs, operational costs, life expectancy, fuel consumption and write-down value – a good introduction to this subject is contained in *The Surveyor's Guide to Civil Engineering Plant*.[4] The *Caterpillar Performance Handbook*[7] also contains a wealth of useful information including a section on building up an hourly rate from first principles, as does F. C. Harris's book *Modern Construction and Ground Engineering Equipment and Methods*.[8]

A further possible useful source of information on rates for construction equipment is the Construction Plant-hire Association schedule of rates which is available to members of the association. The monthly magazine *Plant Managers Journal* also publishes current rates for construction equipment.

Quotations for varied work

Both the New Engineering Contract and GC/Works/1 (edition 3) formally introduce the idea of the contractor being required to submit quotations for varied work before the work is executed. Such quotations should include the full cost of the variation including all the possible costs of delay and disruption. This approach is not unusual on major projects.

The author has a vivid recollection of a contractor's project manager giving to the resident engineer a verbal quotation of £300 000 for a major variation. After the initial rejection the contractor was told to go away, think about it and submit a written estimate which he duly did after considering all the possible disruptive effects – in the sum of £400 000!

After due consideration and consultation with the employer the resident engineer considered the price quoted too high and instructed the work to be executed on a daywork basis. When the work was complete the resident engineer was then fully able to justify to any auditors the total amount of the variation – all £500 000 of it! Unfortunately not only did the exercise cost the employer more than it might have done but the contractor had no incentive to complete early and the progress of the works was adversely affected.

The moral of the story from a client's viewpoint: 'Accept it quick, accept it cheap.' Contractors beware!

Items for consideration when valuing variations

Many factors should be considered when assessing the valuation of varied works, whether the price is established before or after the work is executed including:

1. *General items* – revised method statement, effect on critical path, revised production rates, out of sequence working, restricted site access, summer to winter working, changed nature of ground, late payment and financing costs, etc.
2. *Labour* – uneconomic working, attraction money, additional bonus, overtime payments, shift work, accommodation and welfare, etc.
3. *Construction equipment* – plant available on site, additional mobilisation/demobilisation, additional scaffolding, hoisting or cranage, replanning, repositioning, standing charges, working out of sequence, etc.
4. *Materials* – late orders, additional procurement costs, airfreight, premium costs, small quantities, excessive waste, additional testing, etc.

Completion of new rate form

When agreement has been reached with the contractor the contract administrator should complete a 'new rate form' which will contain a detailed build-up of the rate. Upon completion this should be presented to the engineer/architect/project manager for scrutiny and signature. After signature the contractor should be notified in writing of the amount at which the rate has been agreed.

'New rate forms' should be completed for all rates in the final account which are different from those contained in the B of Q; this is particularly important in the case of major public works which may be subject to detailed audit.

Legal cases of interest

There have been many cases involving variations which have gone through the courts; the following are of particular interest.

Simplex Concrete Piles Ltd v. St Pancras Borough Council (1958)
Simplex undertook to design, carry out and test the piling for the foundation for a block of flats. Conditions made it impossible to carry out the work as tendered and the contractor offered two alternatives. The architect accepted the second alternative which involved using a bored piling subcontractor.

It was held, under this RIBA contract, that the architect's letter was an architect's instruction involving a variation which entitled the contractor to payment.

Dudley Corporation v. Parsons and Morrin Ltd (1959)
In this case the contractor had mistakenly priced an item of excavating 750 cube yards in rock at £75, i.e. 2s. (10p) per cube yard instead of a more realistic rate of £2 per cube. In the event 2230 cube yards of rock were encountered. It was held that the contractor was entitled to the rate of only 2s. per cube for the whole of the quantity of rock excavated.

This judgment under an old edition of the RIBA quantities form is considered today by some observers to be controversial and may not be appropriate in every case. The author recollects a general observation made by John Uff QC at a conference on civil engineering claims:[9] 'Don't ever think what is in textbooks is the definitive answer!'

A. E. Farr Ltd v. Ministry of Transport (1965)
This case, which involved a measurement problem, was based on the standard ICE Conditions of Contract with a B of Q. In accordance with the Conditions of Contract the rates and prices were to cover all the contractor's obligations under the contract necessary for the proper completion of the works.

Clause 16 of the B of Q stated that the measurement of pits and trenches was based on the net plan area multiplied by the depth with working space paid for as a separate item. In the event no separate item for working space was included in the B of Q.

The contractor won his case when the House of Lords held that there was a promise to pay for whatever working space might be necessary whether or not described as a separate item in the B of Q.

It is interesting to note that Duncan Wallace, the author of *Hudson's*, considers that the decision in this case was incorrect but Emden's *Building Contracts* (8th edition, volume 1, p. 143) takes an opposite view!

Mitsui Construction Company Ltd v. The Attorney General of Hong Kong (1986)

The conditions of contract were specially produced by the Hong Kong government and were based on the ICE 4th Edition and JCT 63. The 2-year project involved the construction of a tunnel 3227 m long and 3.6 m diameter. The ground conditions were extremely variable and the engineer specified five different types of tunnel lining suitable for the different ground conditions.

In the event the ground conditions were far worse than anticipated and the contractor was required to construct far more of the heavily designed tunnel section (2448 m compared to 275 m billed) and much less of the unlined section. An extension of time of over 2 years was granted.

The engineer argued that the changes in quantity were not the result of an official variation order and the contractor was paid at the rates in the B of Q. The contractor claimed that the increased quantities amounted to variations and that revised rates should apply. The government took the view that the engineer had no power to revise the rates.

The Privy Council took 'a sensible and business like approach' and found in favour of the contractor, stating that the engineer did indeed have the power to fix a revised rate. This case would seem particularly relevant in interpreting clauses 51(4) and 52(2) in the ICE Conditions of Contract (6th Edition).

English Industrial Estates Corporation v. Kier Construction Ltd (1991)

Two contracts were let to Kier in June and November 1987 for the reclamation at the former Dunlop factory at Speke in Liverpool. The excavation for both subcontracts was sublet to J & B Excavation Ltd.

In the specification the contractor was given a choice for structural fill of either using the material arising from the demolition or importation.

The contract required the contractor to submit his programme and method statement with his tender. The contractor's method statement showed that the excavation subcontractor intended to crush *only suitable* material arising from demolition with the remainder removed from site.

In January 1988 the engineer wrote to the contractor instructing him to crush *all* hard material arising from the site. The subcontractor claimed compensation for the losses due to the additional costs involved.

The arbitrator held that the contractor's method statement was a contract document and the engineer's instruction was thus a variation under clause 51 of the ICE 5th.

On appeal the High Court agreed with the arbitrator's decision.

Example No. 2 – using B of Q rate to establish new rate

The project involves the construction of a reinforced concrete pumping station 40 × 40 m on plan with 3.5 m below ground on a greenfield site.
Further relevant information:

1. ICE 6th and CESMM3 apply.
2. The contractor's method statement (clause 14) indicates a battered sides and backfill approach.
3. The soil investigation reports indicate a band of shale to a depth of 6 m with the water table level at 6 m below original ground level.

A variation order covering the issue of a revised drawing indicates the depth of the excavation increased from 3.5 to 4.75 m.

Under clause 52(2) of ICE 6 the contractor considers that due to the variation the rate for excavation in the B of Q is rendered unreasonable. In accordance with clause 52(2) the contractor should give notice to the engineer, before the varied work is commenced or as soon thereafter as is reasonable in all the circumstances, of his intention to claim a varied or 'star' rate.

The item in the B of Q reads as follows:

CLASS E: EARTHWORKS
E425   General excavation,
        maximum depth 2–5 m   5600 m³   £6.89   £38 584.00

Points to note in CESMM3:

1. Definition rule D1 states 'Excavated material shall be deemed to be material other than topsoil, rock or artificial hard material unless otherwise stated in item descriptions.'
2. Coverage rule C1 states that 'Items for excavation shall be deemed to include holding sides of excavation, additional excavation to provide working space and removal of existing services.'
3. Disposal of 5600 m³ of excavated material is measured as a separate item (E532). (The contractor's estimator should include in the disposal rate for an increased quantity due to bulking.)

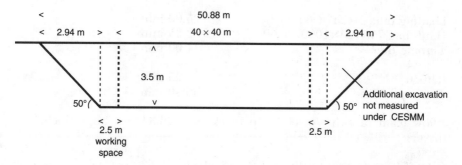

Figure 11.3   Contractor's proposed method of construction at tender stage

An examination of the contractor's estimate shows that he has based his bid on the following:

1. An excavation with a safe slope of 50 degrees from the horizontal.
2. A working space 2.5 m wide around the pump station.
3. A compaction factor of 1.0.
4. Removal of additional excavated material to a temporary stockpile on site 750 m distant from the pump station as required by the specification.

Volume of excavation as B of Q 40×40×3.5 m = **5600 m³**
Actual volume to be excavated:

Main volume: 45 × 45 × 3.5	7087.50
Sides: 4 × (45 + 50.88)/2 × 0.5 × 3.5 × 2.94	986.61
	8074.11 m³

Volume requiring temporary stockpiling and backfilling around pump station on completion of structure:
8074 − 5600 = **2474 m³**

Calculation of rate in B of Q rate – as contractor's estimate sheets

Use a plant and labour gang consisting of:
1 Hymac excavator @ £26.60 per hour (includes driver)
1 ganger @ £6.91 per hour
2 labourers @ £6.55 per hour
Total cost of gang £26.60 + £6.91 + £13.10 = £46.61

Assume output rate of 24 m³ per hour

Cost of excavation £46.61/24 = £1.94 per m³

Excavated material to be transported 750 m to temporary spoil heap using Volvo articulated dump trucks transporting 10.0 m³ per cycle assuming an average haul speed of 20 km per hour.

Loading time 10/24 × 60	25.00 min
Haul time 750/(20×1000) × 60	2.25 min
Tipping time	1.00 min
Return	2.25 min
Total cycle time	30.50 min

Allow team of two Volvo dump trucks @ £26.34 per hour (including driver)

Cost of transport and offload 2 × £26.34/24 = £2.20 per m³

Summary of tender rate build-up

Excavation	8074 m³ × £1.94	£15 663.56
Transport to stockpile	2474 m³ × £2.20	£5 442.80
Excavate from stockpile, return and deposit around pump station	2474 m³ × £3.75	£9 277.50

(based on excavation output of 30 m³ per hour)
(£46.61/30 = £1.55 + £2.20 = £3.75)

Consolidation of backfill	2474 m³ × £0.20	£494.80
Allowance for pumping		£500.00

	£31 378.66
Site overheads + 12 per cent	£3 765.44
	£35 144.10
Head office overheads and profit + 8 per cent	£2 811.53
	£37 955.63
Divide by 5600 m³	£6.78 per m³

Bill of quantities rate = £6.78 per m³

N.B. All off-site disposal, including allowance for bulking, measured and priced as a separate item in the B of Q.

Revised volume of excavation as remeasured in accordance with CESMM3:
40×40×4.75 = **7600 m³**

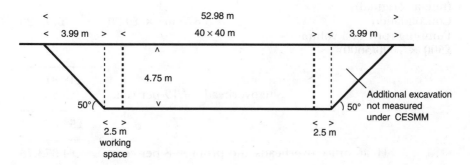

Figure 11.4   Details of the contractor's 'as-built' method of construction after variation increasing depth to 4.75 m

Actual volume to be excavated:
Main volume: 45 × 45 × 4.75   9 618.75
Sides: 4 × (45 + 52.98)/2 × 0.5 × 4.75 × 3.99   1 856.97

                                                                    11 475.72 m³

Volume to be transported to temporary stockpiles:
11 476 − 7600 = **3876 m³**

Build-up of 'star rate' for excavation

Assume output rate reduced overall from 24 to 22.5 m³ per hour due to increased depth

Cost of excavation £46.61/22.5 = £2.07 per m³

Loading time 10/22.5 × 60	26.67 min
Haul time	2.25 min
Tipping time	£ 1.00 min
Return	2.25 min
Total cycle time	32.17 min

Allow two Volvo articulated dump trucks @ £26.34 per hour (including driver)
Cost of transport and deposit 2 × £26.34/22.5 = £2.34 per m³

Summary of build-up of 'star rate'

Excavation	11 476 m³ × £2.07	£23 755.32
Transport to stockpile	3 876 m³ × £2.34	£9 069.85
Excavate from stockpile, transport and deposit (rate as original)	3 876 m³ × £3.75	£14 535.00
Consolidation	3 876 m³ × £0.20	£775.20
Pumping (pro rata volumes – £500 × 7600/5600)		£678.57
		£48 813.94
Site overheads + 12 per cent		£5 857.67
		£54 671.61
Head office overheads and profit + 8 per cent		£4 373.73
		£59 045.34
Divide by 7600 m³		£7.77 per m³

'Star rate' for excavation = £7.77 per m³

Premium on rate claimed £7.77 − £6.78 = £0.99 per m³

Additional amount to be reimbursed: 7600 m³ @ £0.99 = £7524.00

The author accepts that this may not be the definitive solution, indeed this may be the starting-point for negotiation between the parties. Further detailed investigation particularly may be required into those items identified under the previous subheading 'Items for consideration'. However, some degree of comfort is afforded by the observation that Charles Haswell's 10-page article on rate fixing in 1963[5] generated 39 pages of discussion!

Tutorial questions

1. What are the most common methods used when evaluating variations in construction projects? What points should be taken into account when evaluating variations for both budget cost control and for incorporation within the final account? (RICS Direct Membership Examination, Project Cost Management, 1986)
2. The rules for the valuation of variations applying to work carried out under JCT 80 may be considered to be inadequate and to be the primary cause of many disputes.
    Discuss techniques and procedures which may be adopted by site management as an aid to establishing cost and gaining adequate financial reimbursement. (CIOB Member Examination Paper II 1987 – Contract Administration)
3. In appraising a number of tenders for a large civil engineering project, it is found that one contractor has priced items at rates much lower than his competitors, and lower than charges obtainable from plant hire contractors local to the intended project. This anomaly is queried but the contractor explains that the rates are in respect of his own plant and he is happy to perform the contract at the rates charged. On this understanding the contract is duly awarded.
    During the performance of the works, delays and disruptions occur and a considerable amount of work is carried out additional to the amount originally envisaged. In due course an extension to the contract period is awarded which equates to double the period originally intended for carrying out the works.
    During the course of the contract, the contractor indicates that he intends to claim higher rates for plant, saying the work is vastly different, both in scope and time, from that intended, and therefore the original plant rates cannot apply.
    Based solely on the above information, write a detailed report giving your considered opinion as to the validity or otherwise of the contractor's case and comment on all aspects you feel should have been covered prior to awarding the contract and subsequently. (RICS Direct Membership Examination 1986, Quantity Surveying – Project Cost Management)

4. 'Dayworks: a licence to print money or a secure mechanism for correct reimbursement?' Discuss. (RICS Final Examination 1990, Quantity Surveying, – Project Practice and Management, Paper 5)
5. Explain how the engineer responsible for the administration of a contract under the ICE Conditions of Contract would determine the value of payments to a contractor in the following cases:
   (a) where it is decided to carry out a certain section of the work using sulphate-resisting cement when the contract documents envisage all the work being carried out using ordinary Portland cement;
   (b) where the resident engineer requires the use of a lorry and the services of a driver and two labourers to assist in locating and transporting to site various items from the employer's stores;
   (c) where it is decided prior to commencement of excavation, to raise the foundation level of a structure such that the volume of excavation is reduced to approximately 50 per cent of that envisaged in the contract documents.

   (ICE Civil Engineering Law and Contract Procedure, Paper 3, 1986)
6. A contract contains a large earthmoving element including excavating waste tips one of which, when opened, is found to consist of very hot material which gives off dense smoke and fumes. The engineer issues a variation order and the contractor brings on the site a 110 RB excavator and ten 50 tonne dump wagons to deal with the tip. Drivers and plant operators agree to work for a rate equal to double time and are issued with special respirator-type masks. The quantity to be excavated is 350 000 tonnes.

   As the engineer you have to negotiate if necessary and fix a rate for the work. Describe in detail your assessment, making any necessary assumptions regarding plant and labour rates and overheads. Under which clause(s) would you deal with the variation? Would you allow anything for profit? (ICE Civil Engineering Law, Paper 2, 1975)

The ICE holds annual examinations on civil engineering law and contract procedure. The examinations are normally held in June, at various centres in the UK and overseas, and can be taken by members and non-members of the institution.

For further details please contact: The Arbitration Office, The Institution of Civil Engineers, Great George Street, Westminster, London SW1P 3AA, UK. Tel. 071-222 7722, Fax 071-222 7500.

References

1. RICS 1975 *Definition of Prime Cost of Daywork Carried out under a Building Contract* 2nd edn RICS Books
2. RICS 1990 *Schedules of Basic Plant Charges for Use in Connection with Dayworks under a Building Contract* RICS Books
3. FCEC 1990 *Schedules of Dayworks Carried out Incidental to Contract Work*
4. ICES 1991 *The Surveyor's Guide to Civil Engineering Plant: a Guide to Current and Former Models of Construction Plant*

5. Haswell C K 1963 Rate fixing in civil engineering contracts, No. 662, *Proc. Instn Civ. Engrs* **24** (February): 223–34 (Discussion Paper: **27:** 192–231)
6. Abrahamson M W 1979 *Engineering Law and the ICE Contracts* 4th edn Elsevier Applied Science Publishers
7. The Caterpillar Tractor Co. *Caterpillar Performance Handbook* (UK agents: Finning 0543 462551)
8. Harris F C 1994 *Modern Construction and Ground Engineering Equipment and Methods* 2nd edn Longman Scientific & Technical
9. 'Civil Engineering Contracts and Claims' conference Friday 27 November 1987, The Portman Hotel, London W1, Legal Studies and Services Limited

Further reading

CTA Services (Swindon) Ltd 1990 Distance Learning Course, Civil Engineering Law and Contract Procedure

Davis, Langdon, Everest (eds) 1993 *Spon's Civil Engineering and Highway Works Price Book*

Duncan Wallace I N 1970 *Hudson's Building and Engineering Contracts,* 10th edn (First supplement 1979), Sweet and Maxwell

Knowles R 1987 An oriental problem. *Chartered Quantity Surveyor* July

McCaffer R, Baldwin A 1991 *Estimating and Tendering for Civil Engineering Works* 2nd edn BSP Professional Books

Powell-Smith V 1990 *Problems in Construction Claims* BSP Professional Books

Introduction

Claims are inevitable on major construction projects and are usually motivated by a single cause – the contractor spends more money than he expected and he believes someone else is responsible. This additional expense may be due to a single factor or a combination of factors often outside the control of the contractor. Frequently the contractor will be utilising large and costly items of construction equipment and any changes to the works as tendered may cause a delay as well as disrupting his planned production rates with significant financial consequences.

Even though changes are inevitable on major works the additional costs of variations and claims should still be within the promoter's contingency allowance. A vast over-expenditure on claims and threatened or actual litigation can be considered a sign of management failure which may have its roots in the choice of an inappropriate contract strategy.

An improved procedure is indicated in the 'New Engineering Contract' which encourages a less adversarial approach. Under this contract an early warning system stimulates early joint consideration of problems with the estimated total cost being agreed before the commencement of the work.

Types of claims

Contractors' claims fall into three categories:

1. *Contractual claims.* These are claims which can be demonstrated to be due under the contract. The contract will normally require the contractor to serve a written notice with details and substantiation of the claim. The architect/engineer must be satisfied beyond reasonable doubt that the claim is admissible under the actual terms of the relevant clause or clauses of the contract before any payment can be made or extension of time awarded.
2. *Extra-contractual claims.* These are the claims which although not admissible under the contract, appear to be an obligation to the employer which the courts might uphold in common law. Such an

obligation will usually be attributable to the employer's action or inaction. Common law damages claims should be agreed between the employer and contractor, failing which the matter would be referred to arbitration or litigation.

3. *Ex gratia claims.* Even though there may be no entitlement to damages for a breach of contract or a tortuous act by the employer, contractors sometimes submit claims requesting *ex gratia* payment. The usual basis of such an application is that the contractor has suffered a substantial loss on the project which cannot be recovered elsewhere; such claims are rarely entertained by employers.

Legal requirements of claims submissions

A contractor's claim submission must fulfil certain legal criteria:

1. The claim must prove that a loss has been suffered.
2. The claim must show that the loss arose as a result of the relevant acts or omissions.
3. The legal quantum of the claim must be established. The measure of damages in common law remain as stated in *Robinson* v. *Harman* (1848): 'The rule of common law is that where a party sustains a loss by reason of a breach of contract, he is so far as money can do it, to be placed in the same situation, with respect to damages, as if the contract had been performed.'

   This case is of particular relevance in assessing delay and disruption claims as it is the contractor's actual cost at the time of the delay which should be considered and not the amount included in the preliminaries at tender.
4. It must be shown that the loss could not have been mitigated by reasonable conduct. A defendant is liable for such part of the plaintiff's loss only as is properly to be regarded as caused by the defendant's breach of duty. The law does not allow a plaintiff to recover damages to compensate him for loss which would not have been suffered if he had taken reasonable steps to mitigate the loss.

   As an illustration, Powell-Smith and Sims in their book *Building Contract Claims*[1] consider the case of construction equipment standing idle as a result of a variation. Under these circumstances the contractor is bound to make endeavours to use the equipment productively elsewhere on site or arrange for its removal.
5. The losses must not be seen to be too remote. The principles concerning a common law damages claim were set down in *Hadley* v. *Baxendale* (1854). Recoverable damages were separated into two distinct categories:
   (a) First rule – 'such losses as may be fairly and reasonably be considered arising naturally, i.e. according to the usual course of things from the breach of contract'; or
   (b) Second rule – 'such as may reasonably be supposed to have been in the contemplation of both parties at the time that they made the contract, as the probable result of breach of it'.

Delay claims

All construction contracts contain provisions for extension of time. These are usually thought to be for the benefit of the contractor in order that he may avoid the payment of liquidated damages.

However, the provisions are also of benefit to the employer for without such extensions of time clauses in the contract the liquidated damages provisions would cease to have any effect if any delay was caused by the employer or his agents. Time for completion would become 'at large' and the contractor would merely be required to complete in a 'reasonable time'.

Under standard contracts the contractor is required to give notice to the architect/engineer if he believes that the works or any section is in delay. Such notices should be given at the earliest opportunity, indeed clause 25.2.1.1 of JCT 80 requires the contractor to look to the future and to give written notice to the architect if he considers that he is likely to be delayed. The notice should contain full and detailed particulars of the justification of the period of extension claimed in order that the claim may be investigated at the time.

The JCT 80 contract is quite specific on the grounds which can entitle a contractor to an extension of time – the event must be one of the fourteen 'relevant events' identified in clause 25.4. In comparison, the grounds which may give rise to an extension of time under the ICE 6th are identified in clause 44(1) and include the somewhat vague phrase 'other special circumstances of any kind whatsoever which may occur'.

Under the ICE 6th the engineer should grant an interim assessment of the delay provided only that he considers that it will actually be required for the completion of the works. However, it is most important that the engineer should award the interim extension of time promptly and not wait until the end of the project. If the contractor employs additional resources to make up lost time only to find on completion that he is awarded an extension of time the employer may be faced with a claim based on a breach of clause 44.

An extension of time does not of itself entitle the contractor to additional payment. The contractor should look elsewhere in the conditions of contract for clauses which may give grounds for recovery of additional expenses e.g. under clause 26 in JCT 80 or clause 13(6) in the ICE 6th.

Generally speaking, the contractor will normally be entitled to additional payment provided only that the delay is due to an act of neglect or omission of the employer or his agent or those for whom he is responsible, or for unforeseen ground conditions – the risk which is carried by the employer. However, the onus is on the contractor to demonstrate the cause and the effect both for extensions of time and additional costs.

Contractor's programme

The contractor will normally be required to produce a construction programme at the commencement of the project and both parties will rely on this programme in attempting to justify an extension of time. However, unless the programme is submitted with the bid it is unlikely to become a contract document and the client's representative will be under no obligation to accept it as the basis for assessment.

The programme should preferably be in network form in order that the logic can be checked and the critical path established. Normally extensions of time would be awarded only for any items which are delayed and are on the critical path. The contractor may have included 'float' within the programme to allow for any time for which he is responsible, e.g. inclement weather. This float should be considered the contractor's own and should not be utilised by the engineer without compensation to the contractor.

Problems can occur when different causes of delay run concurrently; one cause may entitle the contractor to an extension of time and costs (e.g. late instructions) and the other an extension of time only (e.g. adverse weather). As a general guide it is necessary to establish which cause of delay occurred first, whilst this delay is continuing it can be regarded as the overriding delay.

It should be remembered that the assessment of extensions of time is not an exact science – even if the revised completion date has been established using computers and sophisticated project planning software!

The case of *London Borough of Merton* v. *Stanley Hugh Leach* (1985) is significant to both the forward planning and claims settlements because it considered the use of the programme in relation to the serving notification of delays. The essence of the case involved the employer issuing late information to the contractor where the contractor had to serve notice of the dates on which the information was required. The judge held that the symbols on the contractor's initial programme showed these required dates and this was in fact considered good and sufficient notice.

Whilst this is generally accepted by precedence, it should be recognised that the programme can serve as a notice only if the contractor has not deviated from it, and has kept the programme up to date, properly reflecting delays therein. Certainly under the JCT 80 form of contract the contractor would be advised to give separate notification of information required.

Disruption

The concept of delay is readily understood, disruption on the other hand is more complex. Paul Bryant has identified that disruption could be one of three kinds:[2]

1. The work in question takes longer to complete, using the original resources.
2. The work takes the same time because of increased resources.
3. The contract takes the same time to complete, but certain resources are kept on site longer than originally necessary.

Thus it can be seen that the contractor may be entitled to additional costs for disruption even though he has completed the contract on time or within the extended period.

Under the ICE 6th there is no specific provision for the recovery of such costs; any disruption costs must be recovered under clause 52(4). A realistic starting-point for assessment would be the additional costs that an experienced contractor would not have anticipated at tender.

Likewise under the JCT 80 it is a common misconception that delay is a condition precedence to a clause 26 loss and expense claim; this is not the case as careful reading of the conditions will reveal.

The basis of all additional payments is the information available to the contractor at the time of tender, which is up to the time the offer is finally accepted.

Typical contract documents include the conditions of contract, the specification, the site investigation information, B of Q, tender drawings and form of tender with appendices together with any pre-contract 'questions and answers'.

In order to succeed with a claim for disruption (i.e. in effect loss of output or production) it would be necessary for the contractor to prove that the original tender allowances were sufficient for the requirement, that the organisation on site was as efficient as the circumstances permitted and that all possible steps had been taken (by way of reducing labour force, removal of plant, etc.) to mitigate the effect of the delay. Contractors should also remember that an arbitrator may require a complete disclosure of all documents – including the tender build-up.

If the contractor claims the difference between the tendered rates and the actual cost this may be viewed with some suspicion, particularly if the tender was underestimated or if the contractor was working inefficiently. However, the reason for the contractor's inefficiency may have been due to the out-of-sequence hand-to-mouth working as a result of the disruption.

Unit costings based on actual productivity before and after the disruption could be of some help, but as in most claims, negotiation and compromise will be required.

Unforeseen conditions

Major civil engineering works usually include a considerable amount of work below ground and even though there may have been extensive pre-contract site investigation on land or water the unexpected usually occurs – Murphy's law: anything and everything that can go wrong will go wrong!

In order to appreciate fully the enormous impact which these unforeseen conditions may have upon a major project the following case study is offered as an example: The contract required the construction of a £40m. North Sea oil marine terminal on a tidal river on the east coast of the UK. The project involved the construction of eight jetties working off a total of twelve floating barges. The contractor was required to fabricate and drive over five hundred 900 mm diameter hollow steel tube piles 50–60 m long. Due to site restrictions the piles were fabricated on a wharf 8 km down river and transported in sequence as required by river to the site.

Twelve months after commencement the contractor encountered isolated bands of rock, at a comparatively shallow depth, whilst driving the piles. These bands of rock were not indicated on any borehole reports and proved impossible to penetrate using normal pile-driving methods. As a result the contractor was required to design and fabricate special under-reaming heads and mobilise additional drilling equipment. After weeks of trial and error and agonisingly slow progress the piles were finally driven to the required level.

The consequences of these unforeseen ground conditions were significant and resulted in the contractor incurring substantial additional costs through extra temporary works, additional construction plant and supervisory staff, standing time of crane barges and marine plant, loss of productivity, disruption to the fabrication yard, increased winter working and the effects of rampant inflation.

A further consequence which followed was due to the contractor attempting to make up the lost time by acceleration which involved mobilisation of additional crane barges, payment of improved bonus incentives and additional overtime.

This particular contract, even though in the UK, was based on the FIDIC conditions of contract. Under clause 12 of FIDIC and the ICE 6th the contractor may be entitled to additional costs if he encounters physical conditions (other than weather) or artificial obstructions which could not have been foreseen by an experienced contractor and which result in additional costs.

If it is considered by the courts that an experienced contractor could reasonably have foreseen the event then the claim will fail. In the case of *C.J. Pearce & Co Ltd* v. *Hereford Corporation* (1968), the contractor had to excavate under a 100-year-old sewer known to exist at the time of making the contract. The old sewer fractured when the contractor disturbed the surrounding soil. It was held that the condition was reasonably foreseeable and the contractor lost his case.

Variations

Under clause 52(4) of the ICE 6th if the contractor intends to claim a higher rate than one notified to him by the engineer he should give notice to the engineer within 28 days.

The claim may be in relation to one particular variation where the parties have failed to agree the valuation. If the variation involves extra work valued using B of Q rates the contractor may recover, within the rates, sufficient extra revenue to compensate for any additional on-site and head office overheads costs.

However, problems can occur when the extra time and cost required are disproportionate to the value of the variation or if the variation is issued late in the project resulting in a disproportionate amount of additional costs.

A further problem on major projects is the cumulative disruptive effect on progress of numerous variations which can often be difficult to identify and quantify fully.

Claims procedures

Under the ICE 6th after the contractor has notified the engineer of his intention to claim under clause 52(4) he should include a lump sum estimate of the amount in the next interim valuation (clause 60(1) (d)). After giving the notice of claim the contractor should, as soon as possible in accordance with clause 52(4)(d), send to the engineer full and detailed particulars of the amount claimed and the grounds upon which the claim is based.

Claims presentation

In 1980 the Department of Transport and the FCEC published a 'notice for guidance' on good practice in connection with the submission and consideration of contractual claims.[3] The guidance notes suggested that in accordance with the requirements of clause 52 of the ICE Conditions the contractor should give to the engineer as soon as possible his 'notice of claim' which should:

1. Explain the circumstances giving rise to the claim;
2. Explain why the contractor considers the employer to be liable;
3. State the clause(s) under which the claim is made.

The contractor should, as soon as possible, follow up this 'notice of claim' with a detailed 'submission of claim' which should contain the following:

1. A statement of the contractor's contractual reasons for believing that the employer is liable for the extra costs with reference to the clauses under which the claim is made.
2. A statement of the event giving rise to the claim, including the circumstances (or changes thereof) he could not reasonably have foreseen.
3. Copies of all relevant documentation, such as:
   (a) contemporary records substantiating the additional costs as detailed;
   (b) details of his original plans in relation to use of plant, mass haul diagrams involved;
   (c) relevant extracts from tender programme and make-up of major B of Q rates;
   (d) information demonstrating the individual or cumulative effect of site instructions, variation orders and on-costs relating to the claim.
4. A detailed calculation of entitlement claimed, with records and proofs.

These guidance notes would seem to reflect good practice no matter what the project or the conditions of contract.

A contractor's claim should be submitted in a similar form to that required for a 'statement of case' in the courts. It should be self-explanatory, comprehensive and readily understood by someone not connected with the contract. It should contain a title page, an index, recitals of the contract particulars, relevant clauses and reasons for the claim and an evaluation.

Action required by the engineer

As soon as the engineer receives notification of claim he should consider the matter, notwithstanding that it may be deficient in factual, contractual or financial substantiation and certify accordingly. The Department of

Transport encourages the engineer to certify payments on account up to the level that payments have been substantiated to his satisfaction.

On some projects the engineer may require the permission of the employer before certifying additional monies against claims. The ICE 6th now contains clause 2(1)(b) which requires such modifications to the engineer's authority to be stated against item 18 in the appendix to the form of tender.

As soon as the engineer receives the contractor's 'notice of claim' and prior to any certification it is essential that the engineer marshals the facts of the case into a *claims report*, either for his own records or to send to the employer with his recommendation. The engineer's claims report, which initially can be prepared by the engineer's representative or the contract administrator, should be set out in the following format:

1. Introduction;
2. Summary of claim;
3. Contractor's contention of principles (and value when stated);
4. Analysis of principles of claim:
   (a) circumstances giving rise to claim;
   (b) principles on which the claim is considered (including relevant clauses in the conditions of contract);
   (c) description and justification of claim;
   (d) recommendation on principles of claim;
5. Analysis of quantum of claim:
   (a) analysis of contractor's justification;
   (b) apparent discrepancies;
   (c) recommendation for quantifying claim;
6. Appendices:
   (a) index and copies of correspondence;
   (b) tables;
   (c) drawings.

If insufficient details accompany a contractor's brief notification of claim, the initial report will have to be restricted to items 1–3 and maybe 4 above.

Records required

A party to a dispute, particularly one proceeding to arbitration or litigation, is likely to learn one major lesson – the importance of records. On large projects the contractor and the engineer should agree upon a system which enables contemporary records to be kept from the commencement of the project and not piecemeal when a dispute occurs.

In addition to the contract documents the following information can be of use in resolving a dispute:

- programme including any revisions
- method statement and contractor's proposals
- tender borehole reports including any additional reports
- test results, trials and samples

- working drawings, sketches and bending schedules
- variations, site instructions and confirmation of instructions
- correspondence, minutes of meetings agreed as correct by both parties
- weather records – both pre-contract and during the contract
- labour and plant allocations and returns (daily and/or weekly)
- contractor's wage sheets
- invoices from suppliers/subcontractors
- site diaries – both of agent and section engineers
- progress photographs – preferably dated and coinciding with date of interim valuation
- tender build-up
- contractor's internal bonus and cost records
- site surveys and levels
- remeasurement records
- interim valuations and certificates, etc.
- graphs, charts, tables, schedules, calculations
- notices issued in accordance with the conditions of contract
- 'cause and effect' programme(s)

The importance of records, particularly those made at the time of the event and agreed by the parties, cannot be overemphasised.

Generally the site and office records should be impeccable. The site diary, allocation sheets for labour and plant, confirmation of instructions and even verbal comments should all be kept up to date. Programmes and drawing registers should be monitored on a daily basis. The same should also be required of subcontractors.

Potential 'heads of claim'

The objective of all claims is to put the contractor back into the position he would have been in but for the delay or disruption; the original profit (or loss) should remain as included in the bid. It is necessary therefore to consider the actual additional costs incurred by the contractor at the time of the loss provided of course such costs have been reasonably incurred.

The items described under the headings below are frequently encountered as heads of claim:

Site overheads

This is normally the most straightforward item to establish from contemporary records and would often include the following:

1. Supervisory and administrative site staff, i.e. agents, planners, engineers, quantity surveyors, office manager, secretaries, wage clerks, canteen staff, storemen, foremen, plant fitters, non-working gangers, general site labour, etc.
2. Site accommodation including offices, canteen and welfare, sanitary facilities, stores, fabrication yards, construction plant workshops, laboratories, accommodation for employer's representatives including all water, heating, lighting, telephone/fax charges, etc.

3. Construction equipment and tools together with necessary supplies for upkeep, running costs and maintenance, e.g. cranes, hoists, scaffolding, pumps, generators, compressors, weighbridge, wheel washing facilities, Land Rovers, company cars, road sweepers, water and diesel bowsers, personnel carriers and temporary lighting, etc.

Where the construction equipment is hired by the contractor the normal plant rate will be applicable. Where the plant is owned by the contractor it will be necessary for the contractor to prove that he has been prevented from earning profit on other works before a hire rate would be reimbursed. If this proof is not forthcoming a depreciation rate should be calculated; in practice this approximates to two-thirds of the long-term hire rate.

Head office overheads

In principle head office overheads or establishment charges are recoverable; the difficulty is in the ascertainment. There are several theories and methods of evaluation:

Hudson formula
The formula appears on p. 599 of *Hudson's Building and Engineering Contracts:*

$$\frac{h}{100} \times \frac{c}{cp} \times pd$$

where h is the head office overheads and profit included in the contract, c the contract sum, cp the contract period in weeks and pd the period of delay in weeks.

Emden formula
An alternative formula is produced in *Emden's Building Contracts and Practice*, 8th edn, Volume 2, p. N/46:

$$\frac{h}{100} \times \frac{c}{cp} \times pd$$

where h is the head office percentage arrived at by dividing the total overhead cost and profit of the contractor's organisation as a whole by the total turnover, c the contract sum, cp the contract period in weeks and pd the delay period in weeks.

The difference between the Hudson and Emden formulae is the head office percentage. Hudson includes the head office overheads and profit in the contract sum whereas Emden uses the head office percentage for the company during the year during which the matters giving rise to the claim occur.

Eichleay formula

This formula is the best known and most widely used in the USA. The formula computes the daily amount of overhead that the contractor would have charged to the contract had there been no delay. The contractor is then reimbursed that amount of overhead for each day of delay that occurred. The formula is developed in three stages:

Stage 1:

$$\frac{\text{contract billings}}{\substack{\text{total billings for} \\ \text{actual contract} \\ \text{period}}} \times \substack{\text{total overhead incurred} \\ \text{during contract period}} = \substack{\text{overhead} \\ \text{allocatable} \\ \text{to the contract}}$$

Stage 2:

$$\frac{\substack{\text{allocatable} \\ \text{overhead}}}{\substack{\text{actual days} \\ \text{of contract} \\ \text{performance}}} = \substack{\text{overhead allocatable to} \\ \text{contract per day}}$$

Stage 3:

$$\substack{\text{overhead allocatable} \\ \text{to contract per day}} \times \substack{\text{number of days of} \\ \text{compensable delay}} = \substack{\text{unabsorbed} \\ \text{overhead}}$$

N.B. Contract billings = contract sum, total billings = company turnover.

Formula methods have been accepted by the courts when establishing the amount for head office overheads in claims; however, the contractor should be able to demonstrate that the resources could have been readily employed elsewhere.

Doubt has been thrown on the use of formula methods by the judgment in the case of *Tate & Lyle Food Distribution Ltd* v. *GLC* (1982). In this case the judge refused to accept a percentage to cover managerial time and said that actual records were required. However, Crowter[4] points out that the case was not a construction case, was founded on tort and not contract and the overheads were expressed as a percentage of the claim. Crowter further argues that contractors should be reimbursed based on the number of days for which the employer is responsible, using a calculation essentially the same as the Eichleay formula, with deduction for additional head office overheads earned through additional variations.

The author found in his research[5] that contractors frequently base their claims for head office overheads on a formula method whilst client's representatives often consider the method inappropriate. However it is noted that, for expediency, some employers will accept a formula method.

Interest and financing charges

Further to the cases of *F.G. Minter Ltd* v. *Welsh Health Technical Services Organisation* (1980) and *Rees & Kirby Ltd* v. *Swansea City Council* (1985) it is now evident that contractors can recover as 'cost' or 'direct loss and expense' either:

1. The interest payable on the capital borrowed; or
2. The interest on capital that would otherwise have been invested.

The charges apply to the full amount of the loss, from a mid-point between the commencement of the stoppage to the ending of the stoppage, until the date of certification.

The appropriate rate of interest is normally based on the actual rate paid by the contractor, provided it is not excessive. Financing charges should be calculated using the same rates and methods as fixed by the contractor's bank, e.g. compounding interest at regular intervals.

It is important that the contractor gives notice to the client that financing charges are being incurred. Indeed learned opinion considers that the contractor may lose his right to reimbursement if he does not inform the client.

Increased costs

If the contract is fixed price then the additional cost of carrying out the work later than anticipated due to a delay and disruption is generally recoverable. Normally reimbursement would be on the basis of a known formula, e.g. NEDO or Baxter.

Profit

Loss of profit that the contractor could have earned but for the delay and disruption is an allowable head of claim under the rule in *Hadley* v. *Baxendale* (1854). However, in order to succeed with such a claim the contractor must be able to prove that he has been prevented from earning profit elsewhere, for example he has turned away work due to his involvement in the existing contract.

Loss of productivity/winter working

Inefficient or increased use of labour and plant is an acceptable head of claim. It can be established by comparing the actual production output during the disrupted period with the production levels prior to or after the disruption.

If the contractor is able to demonstrate a link between the delay and winter working any additional costs should be reimbursed provided that due allowance is made for any additional working in spring/summer.

Cost of claim preparation

There is very little case law on this item, however leading commentators seem to concur that the contractor's additional costs in preparing the claim and/or the costs of outside consultants may be recoverable provided that the item has not been claimed elsewhere, e.g. under overheads.

Global claims

Sometimes on major projects the contractor will be unable to identify each separate causative event in a delay and disruption claim due to the complexity of the situation.

The case of *J. Crosby & Sons Ltd* v. *Portland Urban District Council* (1967) was concerned with the construction of a 15-inch trunk water main under the ICE Conditions (4th Edition). The contract had been delayed 46 weeks by a combination of matters, some of which entitled the contractor to extra time and/or money and some of which did not.

The arbitrator whose judgment was upheld by the court held that the contractor was entitled to compensation in respect of 31 weeks of the overall delay and awarded a lump sum by way of compensation.

The principle, which is applicable to claims for extensions of time as well as to claims for additional money, is subject to the following important qualifications:

1. The events which are the subject of the claim must be complex and interact so that it is difficult if not impossible to make an accurate apportionment.
2. There must be no duplication.
3. Any financial claim must exclude profit, if profit is irrecoverable under one or more of the heads underlying the claim.

The more recent case of *Mid Glamorgan County Council* v. *J Devonald Williams and Partners* (1992) reinforced the principle that a composite claim is allowable when the claimed costs arise from various events which have a complex interaction and which render the specific relationship between the event and the time/money consequence impossible or impractical to establish.

However, Mr Justice May sounds a warning in the fifth edition of *Keating on Building Contracts*:[6] 'The danger of advancing a composite claim is that it might fail completely if any significant part of the delay is not established and the court finds no basis for awarding less than the whole. It might also conceivably fail if the court were to find that proper identification of causes and losses could have been made.'

Use of computers

In the past few years there has been an enormous advance in the development of computer technology and project planning software, making these systems particularly relevant in the management of claims.

Whilst it is acknowledged that the initial programme submitted by the contractor is often not a contract document, it can often be considered a valid pre-contract statement providing that it has been checked as being realistic and achievable.

Traditionally, even on major projects, this programme would often be in the form of a bar chart. However there is a continuing trend towards the use of networks, e.g. the new GC/Works/1 (Edition 3) requires the contractor to submit a programme in network format.

The use of network analysis systems of planning and control has many advantages when compared to the Gantt or bar chart approach:

1. It enables the project manager to create a model of the project with a clearly defined logic and critical path. Furthermore a working calendar can be defined together with the input of key resources and the calculation of a predicted cash flow.
2. Once the model is defined the project team can predict the outcome of alternative approaches, e.g. fixed time-scale with increased resources or fixed resources with extended time. The model also enables the team to consider the impact of variations and delays before they occur ('what-if' scenarios).
3. The model can be useful in claims management particularly when identifying the effects of delays and disruption.

   David Carrick identified that the original network can be replicated three times:[7]

   Firstly, reflecting those delays which carry time and money, e.g. late instructions.

   Secondly, with delays which carry only time, e.g. exceptionally inclement weather.

   Thirdly, delays which are the sole responsibility of the contractor, e.g. insolvency of subcontractors.

   The end dates of the three programmes can then be compared and the valid extension of time established together with the period of reimbursement for time-related costs and the period of liquidated damages, if any.

It was in 1983 that many quantity surveyors in the UK were first made aware of the power of computerised project planning software by Gordon Hunt who reported in the Chartered Quantity Surveyor:[8]

While I was in America two years ago I saw an estimating system which was eventually linked to a critical path network. The company in question exercised a technique whereby the updated network incorporating the events on the site was fed back through the computer to provide a revised estimate. The cost of the delays/disruption and so on were therefore identified and apparently the difference between the contract estimate and revised estimate was upheld as a valid claim by an American court.

Whilst reflecting a rather contentious and simplistic view of the subject this statement does indicate the enormous potential for the development of computerised project planning and estimating systems in the field of claims management.

Spreadsheets can also prove useful in claims evaluation enabling data to be continuously amended, moved, copied, sorted, abstracted, reformatted, merged with other data and printed complete with graphs and charts.

A well-structured database can also assist in claims evaluation enabling prompt recovery of relevant correspondence, drawing issues, instructions and minutes of meetings.

The introduction of document-scanning equipment has further been a major step forward in handling large amounts of data. Optical character recognition now makes it possible for large amounts of correspondence to be 'read' and stored by computer. Later the computer can be used to search the information for key words or strings of words.

Tutorial questions

1. A building contractor intends to submit to the client a claim for loss and expense because of the late delivery of drawings. Outline the factors you would need to consider in compiling such a claim and discuss what information you would expect to provide in order to support your claim. (RICS Quantity Surveying Final Examination, Law, 1989)
2. As the contractor's project surveyor on a major contract with considerable potential for claims, set out the uses an on-site computer could have in assisting in the preparation and presentation of the eventual claim submission. (RICS Quantity Surveying Direct Membership Examination, Organisation and Management, 1985)
3. It is sometimes suggested that the satisfactory settlement of contractual claims owes more to skills of communication and negotiation than to a detailed knowledge of contractual provisions.

   Discuss with regard to this statement, the factors considered to be important in achieving a satisfactory settlement and comment upon the management support systems which a contractor requires to ensure such an outcome. (CIOB Member Part II, Contract Administration, resit, 1989)
4. You are the engineer on a major contract. Your representative has certified payment in interim valuations of the following claim by the contractor: the cost of using heavy plant already on the site for a large amount of additional works which became necessary during the contract even though these works could have been completed with smaller, less costly plant. The engineer's representative has signed records confirming the use of this plant on the operations. You have reservations about the evaluation of this claim when the final account is presented to you.

   Write a report analysing the contractual situation and identifying the further factual information that you will require from both the contractor and your representative in order to come to a final decision. (ICE Examination in Civil Engineering Law and Contract Procedure, Paper 3, June 1987)
5. Discuss the value to the engineer of the clause 14 programme in the administration of the contract.

   On a major highway project the contractor submitted a programme showing completion in 84 weeks as opposed to the 100 weeks called for in the contract and it was approved by the engineer. Progress was generally somewhat behind programme throughout the contract. The engineer agreed that variations and additional work on the critical path extended the contractor's programme by 5 weeks. The engineer agreed that a general strike in the construction industry had delayed him by a further 2 weeks. The work was completed in 95 weeks and the contractor submitted a claim for the costs incur~~r~~ 11 weeks on site.

Discuss the engineer's response to the claim. (ICE Civil Engineering Law and Contract Procedure, Paper

References

1. Powell-Smith V, Sims J 1989 *Building Contract Claims* 2nd edn BSP Professional Books
2. Bryant P 1981 Delay, disruption and money. *Civil Engineering Surveyor* June: 18–19
3. FCEC 1980 *Notes for Guidance on the Submission and Consideration of Contractual Claims Arising on Department of Transport Trunk Road Contracts* October
4. Crowter H 1989 Head office overheads. *Chartered Quantity Surveyor* February: 28–9
5. Potts K F 1986 Delay and disruption in construction: ascertaining the cost on building and civil projects. M.Sc. project, University of Loughborough
6. May A (ed) 1992 *Keating on Building Contracts* 5th edn Sweet & Maxwell p 216
7. Carrick D 1989 *Networks and the 'what if' scenario. Civil Engineering Surveyor* October: 4–6
8. Hunt G 1983 Networking critical paths. *Chartered Quantity Surveyor* September: 47

Further reading

Barnes M 1989 Assessing the cost of disturbance. *Civil Engineering Surveyor* June: 21–3

Cree C 1990 Control of claims and dealing with financial disputes. In Barnes M (ed) *Financial Control* Thomas Telford, Ch 7

Hughes G A 1977–78 The anatomy of claims on civil engineering contracts. *ASCE Journal* December 1977: 5–6; February 1978: 24–6

Hughes G A 1981 The claims barrier or how to deal with contractors' claims. *Civil Engineering Surveyor* October: 6,7,23; November: 20–3

Keane J 1988 Extensions of time (Parts 1 and 2). *Chartered Quantity Surveyor* August: 13; September: 32

Knowles R 1983 Interest – is it time to grasp the nettle? *Chartered Quantity Surveyor* May: 385–6

Knowles R 1985 Calculating office overheads. *Chartered Quantity Surveyor* December: 207

Knowles R 1992 *Claims: their Mysteries Unravelled. An Introduction to Claimsmanship for Contractors and Subcontractors* 2nd edn Knowles Publications

Legal Studies and Services Limited 1987 Civil Engineering Contracts and Claims, Conference Documentation, 27 November

Midland Studies Centre 1988 ICE Conditions of Contract, Conference Documentation, 13 April

Thomas R 1993 *Construction Contract Claims* Macmillan
Tudor J 1993 The use of the computer in the preparation, negotiation and settlement of construction claims. Dissertation as partial fulfilment of B.Sc. (Hons) Quantity Surveying degree, University of Wolverhampton

Section F

Construction Business Management

13

Information management

Introduction

The construction process relies on good communication at all times. The project manager must ensure that people perform tasks at the appropriate time. For a person to perform he must be supplied with the correct information to allow him to accomplish his task.

The construction industry has been traditionally conservative in its adoption of new working practices but it is now evident that the use of information technology, or 'IT' is essential for reasons of speed and economy of labour.

This chapter examines how information management can be used on major projects.

Standard business software

Standard business software is generally of an excellent quality, and can be used in any industry for a variety of uses. Their flexibility tends to reduce their efficiency for specialist construction-related tasks. Even though these programes represents thousands of man-hours of labour they are excellent value for money due to massive international sales.

Operating systems

A computer needs an operating system before any software will work. The operating system performs all the background tasks such as entering information through the keyboard or displaying results on the screen or printer. The choice of operating system is vital because it limits the software that you can use. The most common operating system is MSDOS (MicroSoft Disk Operating-System) that is found on all IBM personal computers, or 'PC', compatible computers. There are tens of millions of these personal computers in existence so there is an enormous library of software that runs under MSDOS. The operating system was originally designed for computers that were very crude by today's standards, so

Microsoft has launched the 'Windows' operating system extension that provides a more sophisticated environment whilst retaining backward compatibility. Nearly all the computer manufacturers make PC-compatible computers: this gives the buyer freedom of choice and has created a buyer's market.

The Apple Macintosh computer family is the only other contender in the personal computer market but is not PC compatible. However, major software packages have Macintosh versions. The Macintosh's original strength of user friendliness has been severely eroded with the arrival of 'Windows' software.

Mini computers use proprietary operating systems which restrict your choice of software and hardware. However, they tend to be far more sophisticated and allow you to perform several tasks at the same time and share common databases. Examples include Digital Equipment Corporation's VMS and Sun System's Solaris. The 'UNIX' operating system is emerging as an industry standard operating system for multi-user computer systems and can be installed on a variety of different hardware.

Word processing

The word processor is one of the simplest programs but is used more than any other. It allows you to enter and edit text and arrange and format it. A typical construction application is the creation of preliminaries or specifications. You can quickly insert, delete or cut and paste sections. The positioning of text is manually carried out and there is limited maths capability. The current market leader is 'Word Perfect', others include 'Microsoft Word' and 'Lotus Ami'.

Desk-top publishing is a superior form of word processor, but the better word processors have similar capabilities.

Spreadsheets

Spreadsheets are often the first application used by professionals. They allow the creation of maths models in a simple rows and columns format. The screen shows the result of 'what-if' calculations very quickly. They are ideal for single-page calculations of a one-off nature: their strength, and weakness, is their flexibility. Construction professionals can spend hours elaborating a simple spreadsheet and then corrupt the logic. A typical application would be a manhole schedule that calculates the length of formwork, area of brickwork and volume of concrete based on the internal dimensions. The current market leader is 'Lotus 1-2-3' with 'Microsoft Excel' and 'Borland Quattro' strong contenders.

Databases

Databases are more difficult to get used to; they are not nearly as intuitive as spreadsheets. However, they can be used to extract or sort data and quite sophisticated applications can be written with them, a typical example would be a B of Q. The market leader is 'Microsoft Foxbase' with 'Microsoft Access' and 'Borland Paradox' as alternatives.

Graphics

There is no point in being technically correct if the recipient of a report does not understand what is conveyed to him. Business graphics programs can create excellent graphics that are attractive and informative. A bar chart or pie chart can simplify a mass of figures. The most common example is the cumulative cash flow graph comparing forecast expenditure against actual valuations. The current market leader is 'Lotus Freelance'; 'Corel Draw' is better but more complicated.

Communications

Communications software is the least glamorous type of software but will become more important in the future. This category includes different programs all allied to communication but performing different tasks.

The simplest type of communication software is modem-controlling software such as Crosstalk. This allows a computer to transfer data to another computer via a modem and a telephone line. The information, which may be a word processing file or a spreadsheet, is converted into sound and then transmitted down a telephone cable and reconverted at the other end. This process is similar to a fax but has the major advantage of providing the original information which can be edited and reprinted as a 'new' original.

'Terminal emulation' programs are the next type of communication software. This allows a personal computer to 'talk' to a mini or mainframe. This should be a simple task but in reality is complicated by the fact that all the mini and mainframe computers are totally incompatible from manufacturer to manufacturer. Many makes cannot even communicate between their own model ranges! Mainframe computers are accessed by 'terminals' which are basically a screen and a keyboard connected to the mainframe. If a personal, or 'stand-alone' computer is connected to the mainframe and terminal emulation software is run, this will allow a personal computer to communicate with its 'host'. Examples of terminal emulation software include 'Reflection' and 'Vterm'.

The last, and newest, type of communication software is electronic mail or work group software. This normally runs on a network of connected personal computers and allows users to transfer messages or files. As an example you may wish to send a memo: manually it would be typed, photocopied, and mailed to the recipients and hopefully read. By electronic mail you can type the memo and then 'post it' on the electronic mail system to either an individual, a group such as 'all senior surveyors' or to all members of the project team. The software will tell each user that mail has arrived and will notify you if it has not been read. The receivers can read the document on the screen and then save to disk, delete it, edit it and return with comments, or print it. This should save paper and time. Another use is for open diaries so that meetings can be arranged by the computer comparing different people's availability. The best-known examples of electronic mail are 'Lotus Notes' and 'CC: Mail'.

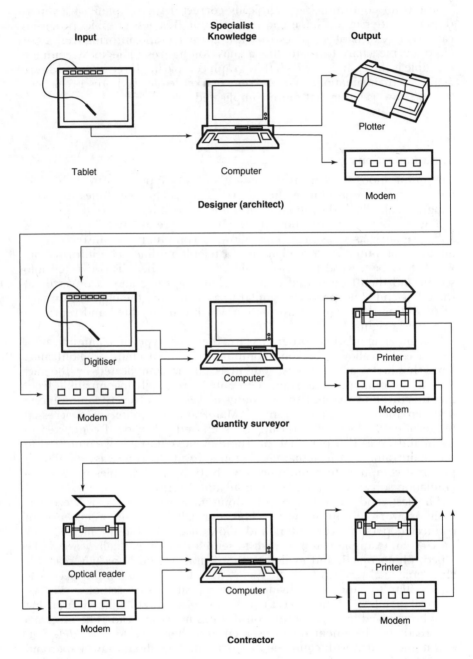

Figure 13.1 Data input and transfer

Construction software

Most software is American and designed for general use, therefore the British construction industry is a very small software market. As a result the software is generally very inferior to the general business software. The better programs are ones that are used in other industries such as engineering.

Project management

Project management (PM) software can be used at many different levels of sophistication for controlling performance. It is most commonly used by the main contractor for ensuring that all the tasks necessary are performed in the correct sequence and in time for the next operation. A consultant can manage his performance, for example B of Q production, to ensure that the individual contributors come together by the finishing date. The overall project manager can use the software to ensure that the client, design team and contractor work together. The principles are simply identifying the tasks, and calculating the duration (or resource expenditure rate) and the relationship between tasks. The software will then calculate the total duration and highlight the critical tasks that have no float. Examples of PM software include 'Artemis', 'Pertmaster', 'Microsoft Project', 'Hornet' and 'Powerproject'.

CAD: computer-aided design

Computer-aided design programmes are more commonly used for computer-aided drafting. The drawing is constructed on the screen, basically being lines from one coordinate to another with a colour and a style. Objects can be common to a level and included or excluded to the drawing, for example the layout grid or the electrical outlets. A library of common objects can be built up, such as cars, trees, toilets, etc., and pasted on to the drawing. The great advantage is the ease of editing and the running off of new originals. Most programs will work in either 2D or 3D modes, the 3D models being the best. If a building is created in 3D then it can be viewed from any angle or sectioned at any point.

In addition it is possible to shade the planes with brick effect or whatever to produce artist's impressions and even to be able to 'walk through' the building on the screen. 3D programs are essential for services drawings to mitigate against collisions and obstructions. Examples include 'Intergraph', mainly found on mini computers, and 'Autocad' which has monopolised the PC market.

Bill of quantities

There are several B of Q programs on the market designed specifically for the British construction industry. In essence they are simple: you select a description from a standard library which has an ordering code, attach a quantity, sort the items and print them. You may then price the items

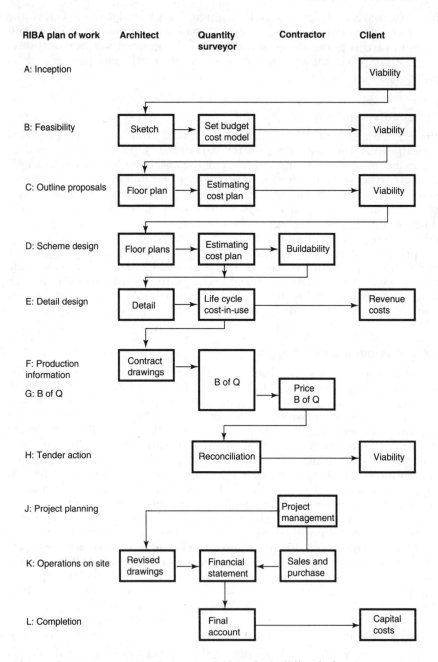

Figure 13.2   Software used in the building industry

and the program will extend the quantity and rate and collect subtotals to a grand summary. The main problem is the enormous number of permutations of items and specifications that can occur which inevitably forces time-consuming creation of 'rogue', i.e. non-standard, descriptions. Examples of B of Q software include 'CATO' and 'Masterbill'.

Other software packages

Many other programs exist for more specific tasks, the majority being in-house applications written using a database or spreadsheet package. These programs tend to be custom produced for special clients. Examples include cost plans, cash flows, and financial statements.

Hardware

Hardware compatibility is generally no longer a major problem. This is principally due to the stabilising influence of the IBM compatible PC. The quality of software has improved due to the vast market for PCs. Unfortunately the PC started as a very humble computer and has been improved over the past 10 years by 'bolting on' improvements. The original $5\frac{1}{4}$ inch disk drives have been replaced by $3\frac{1}{2}$ inch drives. These disk drives save information in one of five different formats. The screen has been improved from the mono text only screen to CGA (colour graphics adaptor), EGA (enhanced graphics adaptor), VGA (video graphics adaptor) and superVGA graphics versions. The speed of the computer has increased with five versions of the central processor.

In addition to the combination of computer options the capability of the hardware is determined by the printer. The output is determined by the limitations of the printer such as whether it can print in colour, text fonts available, graphics capability, etc.

Integration

When you combine the permutations of computers, disk drive size, disk format, printers, operating systems, software types and versions you have a recipe for disaster. If you do not match all the above you will probably have difficulty in being able to read, edit and print the information. This is often a problem within one organisation; the moment you try to communicate with another company you will have problems.

The simple solution is to have exactly the same set-ups but this is obviously difficult to agree on and can be disruptive. Most companies work with many clients and professionals so it is impossible to match all the other parties. There is a shake-out under way within the hardware and software manufacturers which should alleviate this problem in the long term. In the short term the only solution is to try to integrate different systems as much as possible.

File importing and exporting

A typical example of the way in which integration can be achieved was accomplished between a client's mainframe and Bucknall Austin's PCs. The Hewlett Packard mainframe used a proprietary operating system. The site was 80 km away and linked by a megastream dedicated data line, and the mainframe was accessed by Hewlett Packard 'dumb' terminals. The mainframe database was written in a fourth-generation language called 'Speedware' which is not available on PCs.

The first stage was to connect the PC physically to the mainframe by connecting a cable from the serial port to the network outlet. A terminal emulation program called 'Reflection' was installed and set up to emulate an HP terminal. The same software could be used to emulate a terminal for a DEC or IBM mainframe. A log on and password was supplied to allow a user to access the mainframe and run the software. A program was used to search the very large database and select information for the

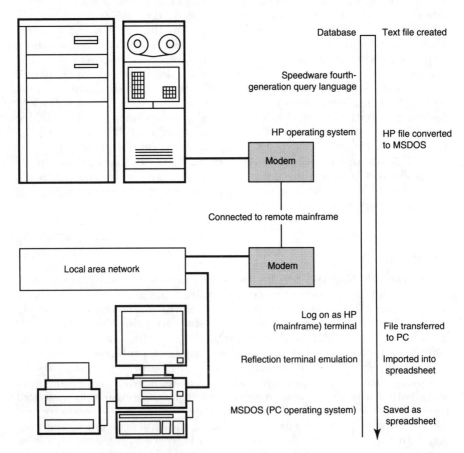

Figure 13.3 Terminal emulation

project and create a file. This 'comma separated file' was then transferred and translated from the mainframe 80 km away on to the PC. The next stage was to log off from the mainframe and enter the spreadsheet 'Lotus 1-2-3'. The file that had been downloaded was not a native 1-2-3 file so it had to be imported using the File Import Text command. Once loaded into the program it was analysed from the lines of text into columns of figures and text. A macro then sorted and printed the data into the required format.

The terminal emulation software was used to feed data into the mainframe once so the client was up to date and ensured that the figures were not different through double entry on to two systems. This mega project was so large that the whole database could not fit on to anything other than a mainframe.

Networking

Originally computers were mainframes. These machines were very expensive and worked on the principle of one large central processor and storage that everybody shared. Everybody worked on the same data and used the same programs, the 'data processing' department built its empire and there was no room for innovation. The PC was invented and suddenly computers were cheap and the end user could work in his own way using whatever software he chose. This started as creative anarchy but quickly became

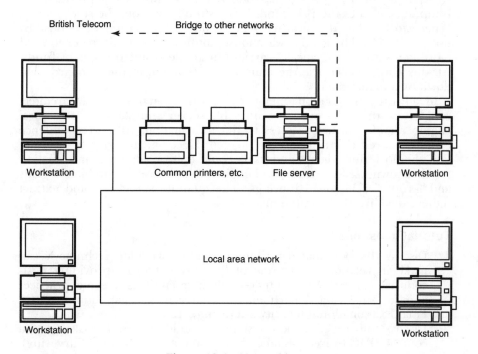

Figure 13.4   Networking

plain anarchy. No data were shared, all information was re-entered on each machine, which is inefficient and technically inconsistent.

Networks were introduced that linked PCs giving them access to shared data but still allowing local processing. They therefore have the advantages of a mainframe with shared data, whilst retaining the creativity of local PCs.

There are two main types of networks, local area networks (LANs) and wide area networks (WANs). LANs consist of one central file server that stores all the common data and a cable running around a building with connection boxes at frequent intervals. An individual PC is connected to the network with its own cable to the socket. On the PC you 'log on' to the network and then you can use the common data or the shared printers. Once information is changed on the network it is available to everyone.

WANs connect LANs in different locations. The most common way of doing this is by having a dedicated data link, connecting the two file servers. This means users at both locations have access to the same information instantly. Digital (DEC) has a world-wide internal network used by over 10 000 people. The technology is in place to link everyone within one company and, by the use of WANs, with other companies as well.

EDI: electronic data interchange

Sainsburys use electronic data interchange now as a commercial necessity At each till on the LAN a barcode reader identifies the product, the computer attaches the price and deducts the item from the store's stock. When stock is low the local file server notifies the central warehouse by use of a WAN. The central warehouse notifies the manufacturer to send more items and automatically carries out an electronic transfer of funds. All this happens because the manufacturers agree to the standards laid down by a major client.

Ideally the same could happen in the construction industry; the architect could store drawings on one central file server and allow the structural engineer and contractor access to the information. All correspondence could be stored electronically and logged. Electronic transfer of information should reduce delays and users could edit or adapt the information for their own use. The structural engineer could use the architect's grids and layout for his own drawings. The quantity surveyor could extract dimensions directly from the files.

Tutorial questions

1. Discuss the potential contribution that information technology can make in relation to the role of quantity surveyors providing post-contract functions for both consultant quantity surveying practices and for surveying departments in contractors' organisations. (RICS Final Examination (Quantity Surveying), 1991)
2. Discuss the use of expert systems in building cost forecasting and control. (RICS Direct Membership Examination (Quantity Surveying), Project Cost Management paper, 1989)

3. Discuss the role of databases in the effective management of an organisation. (RICS Direct Membership Examination (Quantity Surveying), Organisation and Management paper, 1989)
4. As the contractor's project surveyor on a major contract with considerable potential for claims, set out the uses an on-site computer could have in assisting in the preparation and presentation of the eventual claim submission. (RICS Direct Membership Examination (Quantity Surveying), Organisation and Management paper, 1985)
5. Consider the criteria for the selection of computer hardware and software for a medium-sized quantity surveying practice having a varied workload
6. 'Without a clear strategy for computing related to a business plan and a training strategy, it is difficult for an organisation to take optimum advantage of technology to manage information.' (Gordon Kelly, chartered quantity surveyor, July/August 1992)
   Prepare a report (no more than 1000 words) critically appraising your organisation's IT strategy

Further reading

Construction Computing, CIOB, bimonthly
Construction Industry Computing Association, 1991 *Using computers for Project Management*
Construction Industry Computing Association, 1992 *Using Computers for Office Management*
Construction Industry Computing Association, 1993 *Building on IT for Quality*
Kelly G Computing column in *Chartered Quantity Surveyor*, monthly
KPMG, Peat Marwick and the Construction Industry Computing Association 1990 *Building on IT for the 90's*

Section G

Standard Forms of Contract

Brief introduction

The actual forms of contract used on major construction works are as varied as the major project themselves.

A detailed examination of the portfolios of four of the top ten UK building and civil engineering contractors shows the following projects completed/commenced by them in recent years:

- £160m. immersed tube, Conwy, joint venture – two companies, traditional organisational method (ICE 5th);
- £100m. defence facility (floating dock), UK, traditional (GC/Works/1 – edition 2);
- £80m. industrial complex, UK, traditional (JCT 80);
- £84m. British Gas building, Loughborough, management contracting (JCT 87);
- £53m. Glaxo laboratories, UK, design and build (special);
- £92m. water scheme, Nigeria, traditional (FIDIC amended);
- £350m. Tsing Ma bridge, Hong Kong, joint venture, traditional plus contractor's alternatives (special);
- £120m. army camp, Sultanate of Oman, traditional (Omani general conditions based on FIDIC);
- £800m. Chek Lap international airport, Hong Kong, multidisciplinary consortium of multinational contractors – turnkey, lump sum, (special);
- £2000m. Channel Tunnel, part lump sum, part traditional, part design and build (special);
- £370m. power station, Barking, full turnkey in consortium with GEC-Alsthom (modified conditions of contract for the purchase of mechanical or electrical equipment – MF1);
- £400m. nuclear waste disposal plant, Thorp, traditional (GC/Works/1);
- £65m. bridge approach road, Severn Crossing, traditional (modified ICE 5th);
- £50m. Jumbo Jet airport base, Cardiff, traditional with contractor's design portion supplement (JCT 80);
- £96m. hotel, Nassau, design and build (negotiated contract based on I. Chem. Eng. conditions);

- £39m. motorway extension, M8, novated design and build (ICE 5th with Scottish Office major amendments);
- £213m. water tunnels, Lesotho, joint venture, construct and arrange finance (FIDIC 4th);
- £73m. power station, Dubai, turnkey design and construct [based on FIDIC (M&E) and FIDIC (civils) specially amended];
- £83m. hydroelectric project, Sri Lanka, arrange finance and construct (special);
- £35m. airport terminal, Glasgow, design build and manage with guaranteed maximum price (JCT 81 with contractor's design);
- £260m. Second Severn Crossing, finance design build and operate (tailor-made contract);
- £257m. airport terminal, Stansted, construction consultancy and hybrid construction management (special contract);
- £220m. Toyota car factory, Derby, management contracting (JCT 1987);
- £450m. research facility, Stevenage, joint venture with MK-Ferguson, management (tailor-made contract written by client);
- £300m. power station, Sizewell, traditional (client's own conditions based on ICE 5th);
- £178m. hospital, UK, management contracting (special contract);
- £80m. office developments (3), management contracting (based on JCT 1987 with amendments);
- £50m. river barrage, Tees, design and build (FIDIC modified);
- £71m. railway trackwork, Channel Tunnel, subcontract (special sub-contract conditions);
- £95m. (total) defence facilities (2), UK, traditional (GC/Works/1 – edition 2);
- £50m. motorway, M40, traditional (ICE);
- £40m. motorway widening, M40, traditional (ICE);
- £50m. motorway widening (viaduct), M6–Thelwall, traditional (ICE);
- £70m. tunnel, UK, design and build, joint venture – two companies, Medway (ICE modified).

A fifth contractor, whose projects included Canary Wharf and the £500m. city-centre development in Kuala Lumpur, identifies construction management, PM, management contracting and general contracting as the favoured organisational methods on major projects.

A sixth contractor shows a portfolio of management, part design and build and traditional organisational methods with a predominance of 'special' conditions of contract.

In view of the variety of organisational methods and forms of contract used on major construction works the author has decided to review the latest editions of some of the most significant standard forms, viz:

- a traditional form (ICE Conditions of Contract 6th edition);
- a management form (JCT Management Contract 1987);
- an international form (FIDIC 4th edition);
- and a new form designed for use in the UK and overseas which is flexible enough to be used on most major civil engineering or building projects (New Engineering Contract).

Bearing in mind that there are over 30 standard forms of contract currently in use in the UK construction industry the author accepts that this review of the relevant standard forms will not be fully comprehensive. The review, however, should provide a good overview of typical contractual and financial procedures likely to be encountered.

Acknowledgement

The author is most grateful to Costain, Balfour Beatty, Tarmac, Laing, Bovis and Taylor Woodrow for supplying this information.

14

ICE Conditions of Contract (6th edition)

Introduction

Introduced in 1991 the ICE Conditions of Contract (6th edition) is set to become the main standard form of contract for civil engineering projects; the exception being government contracts which are based on the GC/Works/1 form. Where tenders are invited on an international basis, particularly in countries with a 'common law' background, the works are placed using the FIDIC (Fédération Internationale des Ingénieurs-Conseils) contract which is now in its fourth edition.

The ICE 6th was introduced to reflect the changing practices in the industry and new decisions by the courts in the 17 years since the ICE 5th was published. The opportunity was also taken to review all the clauses and the language was simplified.

The ICE 6th does not introduce significant changes in the balance of responsibilities between the parties. The wording of a particular clause may have changed when compared to the ICE 5th, but generally the meaning will not have changed. However, certain important amendments were introduced including:

1. Clarification of the powers and duties of the engineer and his staff in a much extended clause 2.

   It is a requirement of clause 2(1)(b) for those matters for which the engineer must seek the employer's authority to be stated in the appendix to the form of tender. Clause 2(8) requires that the engineer acts impartially in all circumstances, except in connection with those matters referred to by clause 2(1)(b).

   Furthermore under clause 2(2)(a) where the engineer as defined is not a single named chartered engineer, the name of the chartered engineer who will assume the full responsibilities of the engineer under the contract should be notified to the contractor within 7 days of the award of the contract.

2. Under the provision of clause 4(2) the contractor is allowed to sublet work or their design, provided the engineer is notified in writing prior to the subcontractor's entry to site. In contrast, the ICE 5th required the engineer's written consent to any subletting.

Not everyone has accepted this change and some clients and engineers have indicated that they will amend the standard conditions in order to retain a degree of control over the selection of subcontractors.

3. Introduction of clauses 7(6) and 8(2) dealing with the situation where the contractor is responsible for the design of part of the permanent works.

  Under clause 8(2) the contractor is required to use all reasonable skill care and diligence in the design of the works for which he is responsible. However in accordance with clause 7(7) the engineer is still responsible for the integration and co-ordination of the contractor's design with the rest of the works.

4. A requirement in clause 11(1) that the employer makes available to the contractor all information on the nature of the ground and subsoil before the submission of the contractor's tender.

5. All additional costs under the clause 12 'Adverse physical conditions and artificial obstructions' now attract profit. (Under the ICE 5th costs of delay or disruption of working did not attract profit – clause 12(3).)

6. A strict requirement in clauses 14(2) and 14(7) for the engineer to accept or reject the contractor's programme and method statement within 21 days after receipt.

7. A requirement under clause 42 that the employer must give access to the site to enable the contractor to commence and proceed with the works and not merely give possession of the site as required by the fifth edition.

8. A tighter time control on the engineer in connection with the final determination of extensions of time as clause 44(5) (within 14 days of the issue of the certificate of substantial completion of the works or section) and the introduction of the provision for accelerated completion in clause 46(3).

9. Substantial revision to the liquidated damages provision in clause 47 to separate the different situations that can exist particularly when the work is divided into sections or when variations are issued after the contract completion date.

10. New powers enabling the engineer under clause 51 to order variations during the defects correction period and to issue variation orders in order to overcome problems caused by the contractor's default.

11. Introduction in clause 66(5) of a conciliation provision in the event of disputes.

Type of contract

In contrast with JCT 80 there is only one version of the ICE 6th which has been developed and agreed between three parties: the Institution of Civil Engineers, the Association of Consulting Engineers and the Federation of Civil Engineering Contractors.

  The ICE 6th is a measure and value contract – the quantities in the B of Q are approximate and subject to remeasurement. This is in contrast to the normal approach on building contracts (e.g. JCT 80, IFC 84) which are normally lump sum contracts subject to adjustment only for variations, fluctuations and claims.

It is argued that the reason for this difference of approach between building and civil engineering contracts is that there is likely to be far more work of an uncertain nature below ground in civil engineering projects.

Some of the advantages attributed to the admeasurement system of contracting are:

- The contractor is paid for the amount of work he actually does.
- There is freedom for the employer to introduce variations whilst retaining a fair basis for payment.
- Adjudication of tenders is simplified as all tenders are priced on a comparable basis.
- The tenderer is given a clear idea of the scope of the work involved through the B of Q.
- Most contractors/subcontractors in the UK are familiar with this type of contract.

However, see Chapter 7 in which some of the disadvantages in using B of Q are considered in detail.

Definitions

Clause 1 of the ICE 6th contains a particularly useful list of definitions and interpretations.

'Tender total' is defined as the total of the B of Q at the date of award of the contract or in the absence of a B of Q the agreed estimated total value of the works at that date. In contrast the 'contract price' is the sum to be ascertained and paid on completion, i.e. after remeasurement.

'Permanent works' are the works to be constructed and completed in accordance with the contract, i.e. indicated on the engineer's drawings and measured in the B of Q.

'Temporary works' normally are those works which are to be designed and provided by the contractor in order that he can execute the permanent works, e.g. temporary cofferdams, bailey bridges, dewatering systems. However, it is not unknown for the engineer to design some items of temporary works and include these in the B of Q. 'The works' means the permanent works together with the temporary works.

The 'defects correction period' is a new term introduced into the ICE 6th and replaces the rather misleading term 'period of maintenance' included in the ICE 5th.

'Cost' is defined as 'all expenditure properly incurred or to be incurred whether on or off the site including overhead finance and other charges properly allocatable thereto but does not include any allowance for profit'.

Any communications which under the contract are required to be 'in writing' may be handwritten, typewritten or printed and sent by hand, post, telex, cable or facsimile.

The contract documents

Clause 1(1)(e) states that the contract consists of a bundle of documents comprising the:

- *Conditions of contract*: (may include special conditions added for the particular works – see clause 71).
- *Specification*: (on major projects this may include both general and particular specifications).
- *Drawings*: (these are the drawings, normally of the permanent works, which are designed by the engineer).
- *Bill of quantities*: (priced and completed; normally prepared in accordance with the Civil Engineering Method of Measurement 2nd edition (3rd edition) or alternatively the Method of Measurement for Highway Works).
- *Tender*: (see the 'form of tender' on p. 47 of the ICE Conditions of Contract).
- *Written acceptance*: (the acceptance of the contractor's tender by the employer should be unqualified).
- *Contract agreement* (if completed): (see clause 9 and the 'form of agreement' on p. 51 of the ICE 6th).

Clause 5 provides that the documents forming the contract are 'mutually explanatory' of one another and any ambiguities should be explained by the engineer and clarified in writing as an instruction issued in accordance with clause 13. Therefore the usual rule that special conditions prevail over standard provisions does not apply.

Parties to the contract

The employer plays very little part in the running of the contract; his main duties are paying the contractor the sums certified by the engineer and agreeing to matters which require changes to the contract itself.

The employer does, however, have some specific rights and obligations under the contract, for example consenting to the assignment of the contract (clause 3), approving the bond and surety (clause 10) and the responsibility for negligence of his own workpeople or agents in connection with damage to persons and property (clause 22(2).

The employer appoints an engineer who is not a party to the contract but who acts in two separate capacities under it namely:

- representing the employer as his agent;
- as a quasi-arbitrator who must act independently between the employer and the contractor.

The role of the engineer

The engineer derives his power to act for the employer from his contract with the employer for the design and supervision of the works. The roles of the engineer can be summarised as listed in the headings that follow.

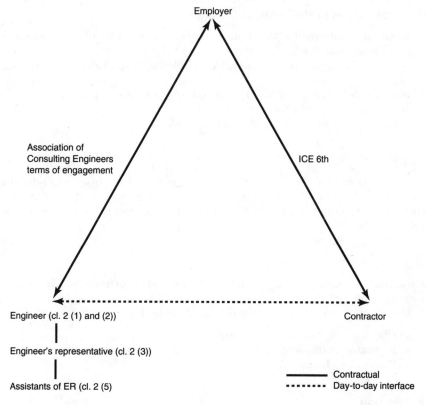

Figure 14.1   ICE 6th: contractual links between the parties

Supervision

The engineer through his representative has to 'watch and supervise the construction and completion of the works'.

Typical clauses in the contract include:

- testing of quality of materials and workmanship (36)
- access to the works, the site and all workshops (37)
- examination of all work before covering up (38)
- removal of unsatisfactory work and materials (39)

Instruction

In accordance with clause 13 the contractor is required to comply with and adhere to the engineer's instructions on any matter connected with the construction of the works whether mentioned in the contract or not. In general, therefore, the engineer can issue instructions about anything provided this is not contrary to the contract.

Typical clauses in the contract include:

- issue instructions to the contractor dealing with adverse physical conditions and artificial obstructions (12(5))
- instruct the contractor to make boreholes or exploratory excavations (18)
- instruct the contractor to provide safety precautions (19(1))
- suspend the works (40(1))
- order variations to the works (51)

Extension

Clause 44 provides that if the work is delayed by causes outside the contractor's control the engineer shall by notice in writing grant the contractor an extension of time for completion (see later detailed examination of clause 44).

Valuation

Clause 56(1) requires the engineer to ascertain and determine by admeasurement the value of the work done.

Other typical clauses referring to valuation include:

- verification of final account (60(4))
- valuation of variations (52(1))
- valuation of claims (52(4))

Certification

The engineer is required to issue numerous certificates, e.g.:

- certificate of substantial completion (48)
- defects correction certificate (61)
- monthly payment certificate (60)

Mediation

Clause 66(3) requires the engineer to decide matters of dispute.

The engineer's representative

Clause 2(3) provides that the functions of the engineer's representative (often referred to as the resident engineer) are to watch and supervise the construction and completion of the works.

The engineer may, under clause 2(4), delegate to the engineer's representative or any other person responsible to the engineer, most of the powers vested in the engineer under the contract. However, such powers will have no effect unless the contractor has been notified in writing of the matters delegated. This delegation could include delegating to the quantity surveyor the duty of remeasurement.

There are, however, certain powers which may not be delegated under the following clauses:

- determination of the cost of delay following the contractor encountering artificial obstructions and adverse physical conditions (12(6));
- award of extension of time (44);
- provision for accelerated completion (46(3));
- issue of the certificate of substantial completion (48);
- certification of the final amount due to the contractor (60(4));
- issue of the defects correction certificate (61);
- certification of the contractor's default (63);
- settlement of disputes (66).

The principal duties of the engineer's representative on a large site include: supervising and checking, carrying out tests, keeping daily records of progress, examining the contractor's programme and methods, checking on safety, measuring the quantities of work carried out, ensuring records kept of additional items and keeping the as-built drawings.

Unless the engineer's representative has been given express authority he cannot order any works which involve delay or extra payment to the contractor.

General obligations of the contractor

Clause 8(1) states that the contractor's general obligation is to construct and complete the works and supply everything that is necessary for completion. Clause 15 further requires the contractor to supervise the works through a competent agent who is normally a chartered civil engineer.

Under clause 8(2) the contractor is not responsible for the design of the permanent works (except as may be expressly provided in the contract) or for any temporary works designed by the engineer. If temporary works are designed by the engineer then the engineer is responsible for their adequacy and the items should be included in the B of Q as 'specified items'.

Clause 8(3) further states that the contractor is fully responsible for the safety of all site operations and methods of construction. Additionally clauses 19 and 20 require the contractor to be fully responsible for the safety of all people on the site and for the care of the works until the issue of the certificate of completion.

In accordance with the requirements of clause 11(1) any ground surveys that are necessary to establish the design criteria should be done by the employer and provided to the contractor at tender stage.

Under clauses 11(2) and 11(3) the contractor is deemed to have based his tender on the information made available by the employer and on his own inspection and examination of the form and nature of the ground and subsoil, the extent and nature of the work and the access to the site. However, the contractor would not normally be expected to carry out additional ground investigation prior to tender.

The case of *Bacal Construction Ltd* v. *Northampton Development Corporation* (1975) involved a dispute as to the accuracy of ground condition information on a design and build contract. It was held that there was an implied term as to the accuracy of the information supplied and the employer was responsible for the cost of the redesigned work.

Clause 12(1) states that the contractor may be entitled to additional costs if he encounters physical conditions (other than weather) or artificial obstructions which could not have been foreseen by an experienced contractor and which result in unforeseen additional costs. Examples of 'physical conditions' i.e. of nature include: natural reservoirs, unpredictable water tables, exceptionally high flood levels, running sand, geological faults including isolated boulders, etc. Examples of 'artificial obstructions' i.e. of man, include: sewers and culverts, services, old structures, old waste tips, etc.

If the contractor intends to claim for additional payment he should give notice to the engineer in accordance with clause 52(4) (claims) and/or clause 44(1) (extension of time). The contractor should also give details of any anticipated effects, the measures he has taken or is proposing to take, the estimated cost and the extent of the delay.

The contractor is required only to construct, complete and maintain the works 'save in so far as it is legally, or physically impossible' (clause 13(1)). The case of *Turriff Ltd* v. *Welsh National Water Development Authority* (1980) concerned the jointing of precast concrete sections of a culvert where the design tolerances made it impossible to join the pipes. The contractor won his case when it was held that he was not liable to redesign the work to render it capable of being constructed.

Programme and methods of construction

Clause 14(1) requires the contractor to submit to the engineer for his approval, within 21 days after the award of the contract, the following documents:

- a programme showing the order of procedure in which he proposes to carry out the Works, and
- a general description of the arrangements and methods of construction which the contractor proposes to adopt.

The programme is an important control document. As a large proportion of work is now subcontracted a contractor's potential ability for making a greater profit by efficient working are somewhat reduced. Therefore the potential savings on overheads by completing the work early has become particularly significant.

Further savings can be effected in a contractor's tender by careful consideration of the method statement and innovative design of temporary works.

As the clause 14 programme is submitted after the acceptance of the tender it is not a contract document. The general position is that the contractor has an obligation to complete the works within the time stated in the contract and it is up to him to arrange his operations to achieve this result.

The programme is required for a number of purposes by all parties.

The engineer will require it to organise his office so that he can supply further drawings and details at the appropriate time. He will also use it to monitor the progress of the contractor and identify any valid extensions of time.

The employer will wish to know the contractor's requirements for site availability, the likely cash flow for the contract and the realistic dates for completion and handover.

The contractor will require the programme to plan and control the works including material purchases, plant mobilisation and organising staff and subcontractors. Contractors will also attempt to use the programme to establish the basis of claims for any delay and disruption.

Clause 14(2) requires the engineer, within 21 days after receipt of the contractor's programme, to accept or reject it stating the reasons, or request the contractor to provide further information. If the engineer does not respond within 21 days he is deemed to have accepted the programme.

Under clause 14(3) the contractor is given 21 days to supply the information requested by the engineer; if he fails to provide such information the programme should be rejected.

Clause 14(4) enables the engineer to request the contractor to produce a revised programme if he considers that the progress of the works does not conform to the programme. In addition, clause 46 gives the engineer the authority to notify the contractor if he considers the progress too slow. This latter clause is important in that it may act as a prelude to the clause 63 forfeiture clause which allows the employer to take over the works.

Clause 14(6) requires the contractor, if so requested by the engineer, to submit details of his methods of construction including temporary works and the use of contractor's equipment. This is to enable the engineer to decide that the methods, if adopted, will have no detrimental effect on the permanent works.

Clause 14(8) provides that if the engineer should delay giving approval to the contractor's methods it may result in the contractor being entitled to an extension of time under clause 44 together with the recovery of the associated costs under clause 52(4).

Legal cases of interest

The case of *Yorkshire Water Authority* v. *Sir Alfred McAlpine & Son (Northern) Ltd* (1985) concerned a dispute on the construction of a £7m. outlet tunnel at Grimworth Reservoir in North Yorkshire under the ICE Conditions of Contract (5th edition).

The contractor was required to submit with his tender a programme showing, *inter alia*, that he had taken note of certain specified phasing requirements – in particular that the upstream work preceded the downstream work. This proved impossible and the contractor proceeded with the downstream work and sought a variation under clause 51(1); 'such changes may include ... changes in the specified sequence method or timing of construction'.

The court held that:

- The method statement was not the programme required to be submitted under clause 14.
- The incorporation of the method statement into the contract imposed an obligation on the contractor to follow it so far as it was legally or physically possible.
- The method statement, therefore, became a specified method of construction and the contractor was entitled to a variation order and payment accordingly.

As commented by Mr Justice Skinner: 'The plaintiff (Yorkshire Water) could have kept the programme and methods as the sole responsibility of the contractor under clauses 14(1) and (3); the risks would then have been the respondents' (McAlpine) throughout.'

The case of *Glenlion Construction Ltd* v. *The Guinness Trust* (1987) concerned a dispute over a residential development at Bromley under JCT 63 (July 1977 revision with quantities). The time for completion in the appendix was 114 weeks and the B of Q contained the provision that within 1 week of the date of possession the contractor should provide a programme chart for the whole of the works. Glenlion submitted a programme showing completion in 101 weeks.

The judge was effectively asked to decide three matters:

- Was the B of Q requirement calling for a programme a term of the contract? (The court held 'yes'.)
- Was the contractor entitled to carry out the works in accordance with the programme and so complete early? (The court held 'yes'.)
- Was there an implied term that the employer and his architect would carry out their obligations to enable the contractor to complete early? (The court held 'no'.)

The judge was influenced by *Hudson's Building Contracts* (10th edition) and *Keatings Building Contracts* (4th edition, First Supplement) which both expressed the view that the architect is under no obligation to issue drawings and details at such times as to enable the contractor to complete to a programme which showed an early completion.

Glenlion lost their case which has considerable implications not only for the JCT form but also for other forms – both standard and non-standard.

Care of the works and insurance

Under clause 20 the contractor is made fully responsible for the care of the works, including unfixed materials, plant and equipment, from the works commencement date until the date of issue of a certificate of substantial completion for the relevant section or the whole works whichever occurs earlier.

However, the contractor is not liable for loss or damage due to an 'excepted risk' which includes damage by the employer or those for whom he is responsible, design error by the engineer, riot or war, radioactivity contamination and sonic bangs, etc. If the contractor should incur any additional expense following the occurrence of an excepted risk he should be reimbursed the costs by the employer.

Clause 21 requires the contractor to insure in the joint names of the employer and the contractor, against any loss as required under clause 20, for the full replacement cost plus an additional 10 per cent to cover for any demolition and professional fees.

It is noted that it is no longer a contractual requirement under clause 21 for contractors to insure their construction equipment; they should of course take out this insurance under a separate arrangement.

Under clause 22, the contractor is required to indemnify the employer against all losses and claims in respect of death or injury to persons or loss or damage to property (other than the works).

However, there are certain exceptions, which are the responsibility of the employer, including: damage due to the construction of the works which is unavoidable, damage to crops or rights of way or damage caused by an act of neglect by the employer or those for whom he is responsible. Clause 22 also provides for an apportionment of the liability depending on the extent of neglect of the other party.

Clause 23 requires the contractor to take out third party insurances to cover the indemnities specified in clause 22. This insurance should cover at least the amount stated in the appendix to the form of tender.

Clause 24 states that the contractor should indemnify the employer against any claims resulting from an accident or injury to any person in the employment of the contractor or his subcontractors.

The contractor should provide evidence that he has taken out the required insurance policies. If the contractor fails to produce such evidence the employer could take out the insurance as necessary and deduct the monies due from the contractor.

Extension of time for completion

Clause 44 (Extension of time for completion) affords possible relief to the contractor by allowing a revised date for completion to be established. The clause is also beneficial to the employer by allowing the retention of the liquidated damages provision. Without such a clause time would become 'at large' and the contractor would be required only to finish within a reasonable period.

Grounds for an extension of time for completion under the ICE 6th are:

- any variation ordered under clause 51(1), or
- increased quantities referred to in clause 51(4), or
- any cause of delay referred to in these conditions: e.g. supply of documents 7(4); unforeseen conditions 12(2); engineer's instructions 13(3); revised method or programme 14(8); public utilities 27(6); facilities for other contractors 31(2); suspension of the works 40(1); failure to give possession 42(3);
- exceptional adverse weather conditions, or
- other special circumstances of any kind whatsoever which may occur.

Clause 44(1) provides that if the contractor considers that any of the above events entitle him to an extension of time he should:

- notify the engineer within 28 days after the cause of the delay, or as soon as is reasonable; and
- submit full and detailed particulars to the engineer.

Clause 44(2) is a new clause and requires the engineer, upon receipt of the contractor's particulars, to make an assessment of the delay that has been suffered by the contractor and notify him of his decision. Furthermore the engineer may make such an assessment even in the absence of the contractor's notification.

Clause 44(3) again is a new provision requiring the engineer to give an extension of time forthwith, but only if he considers that it will actually be required for completion of the works, or any section.

Clause 44(4) requires the engineer to review the contractor's entitlement for an extension of time no later than 14 days after the due date or extended date for completion of the works or any section.

Clause 44(5) requires the engineer to make a final review of the contractor's entitlement within 14 days of the issue of the certificate of substantial completion for the works or of any section.

The granting of an extension of time under clause 44 will not automatically entitle the contractor to recover any additional costs – these should be recovered elsewhere under the contract, e.g. under clause 52(4).

Variations

Because of the uncertainties inherent in the variable nature of the ground and the complexity of the work variations are inevitable on civil engineering works. However, the engineer does not have the power to order variations outside the scope and terms of the contract; if these are required they should be by mutual agreement between the employer and the contractor.

Clause 51(1) requires the engineer (or the engineer's representative) to order variations to any part of the works, that may, in his opinion, be necessary to enable the works to be completed or desirable for the satisfactory completion and/or improved functioning of the works.

Such variations may include additions, omissions, substitutions, alterations, changes in quality, form, character or kind, changes in position dimension or level or changes in any specified sequence method or timing of construction and may be ordered during the defects correction period (clause 51(1)).

No order in writing is necessary for increased or decreased quantities where due to an error in the B of Q and not a variation (clause 51(4)). However, it is noted that clause 56(2) permits the revaluation of any rates rendered unreasonable or inappropriate by changes in quantities.

If a variation renders any rate inappropriate then either party should notify the other, preferably prior to commencement of the work, and the engineer should fix such a rate as considered reasonable and proper.

All variations should be made in writing by the engineer but if made orally they should be confirmed in writing as soon as possible. However, contractors are advised to confirm all engineer's oral instructions. If such instructions are not contradicted forthwith by the engineer then, in accordance with clause 2(6), such an instruction is deemed to be a valid order as if given by the engineer himself.

Variations are valued by the engineer in accordance with the principles stated in clause 52(1) after consultation with the contractor. If the parties cannot agree the engineer should fix the rate and notify the contractor accordingly.

Dayworks

Where the varied work is not a measurable item it may be appropriate for the engineer to order that the work is valued on a daywork basis in accordance with clauses 52(3) and 56(4).

Before ordering the materials for the daywork item the contractor should submit to the engineer, if so required, quotations requiring approval. It is suggested that this procedure would normally apply only in the case of materials with a significant cost and where the work is not urgent.

After executing the work the contractor should record all the labour, materials and plant, etc. involved in the daywork item and deliver such records to the engineer's representative for verification as directed.

At the end of each month the contractor should deliver to the engineer's representative a priced statement of all the dayworks. This statement is normally included in the contractor's monthly interim valuation.

The contractor is paid for dayworks either:

• under the conditions set out in the dayworks schedule in the B of Q, or
• in accordance with the FCEC's *Schedules of Dayworks Carried out Incidental to Contract Work*.

Daywork records are particularly important on civil engineering contracts as a means of recording varied work. Even if the engineer has not ordered that the work be executed on a daywork basis the contractor should ask the engineer's representative to stamp the records as 'agreed for record purposes only'. The priced records may later form the basis of significant claims.

Claims

Under clause 52(4) if the contractor intends:

• to claim a higher rate than the one notified to him by the engineer, for a variation or following a change in actual quantities being greater or lesser than the B of Q, or
• intends to claim any other additional payment pursuant to the conditions of contract,

then he should notify the engineer in writing within 28 days of such notification or as soon as is reasonable. The contractor should keep 'such contemporary records as may be reasonably necessary' to support any such claim, e.g. verified daywork sheets.

The clauses with express reference to clause 52(4) in the ICE 6th and which may give rise to delay and an additional cost claim are:

- documents mutually explanatory (5, 13(3))
- further drawings, specifications and instructions (7(4))
- adverse physical conditions and artificial obstructions (12(2))
- engineer's instructions (13(3))
- revised methods or programme (14(8))
- public utilities – delays attributable to variations (27(6))
- facilities for other contractors (31(2))
- suspension of work (40(1))
- possession of site and access (42(3))

In accordance with clause 60(1) when a claim arises the contractor is required to include in his monthly interim valuation the estimated amounts (i.e. lump sum) to which he considers himself entitled. However, the full particulars required under clause 52(4)(d) normally require considerable preparation and would be submitted at a later date.

Further clauses under which additional payments may arise are: 17, 18, 20(2), 22(2), 26(2), 32, 36(3), 38(2), 49(3), 50, 52(4), 69 and 72.

Measurement

The ICE Conditions of Contract contemplates the use of B of Q and the work being subject to admeasurement. Clause 55 states that the quantities set out in the B of Q are the estimated quantities and are not to be taken as the actual or correct quantities of the work.

The engineer should ensure that the quantities are as accurate as possible. Clause 55(2) requires that any error in description or omission from the B of Q should be corrected by the engineer and treated as variation to be valued in accordance with clause 52.

Under clause 56 the responsibility for the remeasurement and valuation of the works rests with the engineer. However, in practice the remeasurement tends to be a joint exercise; often the contractor's staff measure the works with the checking carried out by the engineer's representative or consultant quantity surveyor.

Clause 57 requires that, unless otherwise provided for in the contract, the B of Q are prepared in accordance with the Civil Engineering Standard Method of Measurement 2nd edition 1985. The appendix to the form of tender allows the engineer to insert an alternative method of measurement, e.g. the Method of Measurement for Highway Works for road and bridge contracts placed by the Ministry of Transport.

Certificates and payments

Clause 60(1) requires the contractor to submit to the engineer at monthly intervals a statement showing the following:

1. The estimated value of the permanent works executed up to the end of that month. The contractor would normally keep the admeasurement up to date and should be able to include reasonably accurate figures each month. The statement should be based on the gross value to date rather than on an incremental basis.
2. A list of materials on site, not yet incorporated into the permanent works, with their value.
3. A list of any off-site goods which have previously been listed in the appendix to the form of tender with their value; these goods must be vested with the employer (e.g. structural steelwork or bridge bearings) (clause 54 identifies the detailed provisions regarding vested goods).
4. The estimated amounts to which the contractor considers himself entitled and for which the contract provides for payment, i.e.:
    (a) temporary works and constructional plant for which separate amounts are included in the B of Q;
    (b) extra cost due to engineer's instructions (clause 13(3));
    (c) payment for fees rates and taxes under clause 26;
    (d) variations under clause 51 and valued under clauses 52(1) and (2);
    (e) dayworks under clause 52(3);
    (f) claims under clause 52(4);
    (g) a separate list of payments to nominated subcontractors under clause 59.

After receipt of the monthly statement, the engineer must certify and the employer must pay within 28 days. If they do not do so then the contractor is entitled to interest on the overdue payments as stated under clause 60(7), i.e. 2 per cent per annum above the base lending rate of the bank specified in the appendix to the form of tender.

Clause 60(3) provides that until substantial completion the engineer is not required to issue an interim payment certificate for a sum less than the minimum value as stated in the appendix.

Final account

In accordance with the provisions of clause 60(4) the contractor is required to submit to the engineer a statement of final account not later than 3 months after the date of the defects correction certificate.

Within 3 months of receipt of the contractor's final account and all verifying information the engineer is required to issue a certificate stating the amount which in his opinion is due to either the contractor or employer.

Retention

A retention percentage is to be inserted in the appendix to the form of tender – the rate is recommended not to exceed 5 per cent of the amount due (excluding materials on site or vested materials). The limit of retention should also be stated in the appendix – recommended limit 3 per cent of the tender total.

After the issue of a certificate of substantial completion for a section of the works or for all the works half the relevant retention should be released and paid to the contractor within 14 days.

The remaining half of the retention is released at the end of the defects correction period (normally 12 months after completion); the employer is, however, permitted to withhold the estimated value of any outstanding work.

Fluctuations

On civil engineering contracts with a construction duration longer than 12 months a contract price fluctuations provision is often included. The operation of this provision, which is known as the 'Baxter formula', is described in detail in the insert found at the back of the ICE Conditions of Contract.

Under the 'Baxter formula' a contractor's bid rates remain unchanged; contractors are not required to prove any increases as the monthly adjustments are based on indices published by HMSO. The procedure broadly is as follows:

1. Prior to inviting tenders the employer (or engineer on his behalf) is required to estimate the proportions of the following included in the works:
    (a) labour and supervision;
    (b) plant and transport;
    (c) materials (in twelve categories).

    The total must represent 90 per cent of the value of the works with a 10 per cent non-adjustment item. These percentages are usually stated in the general items section of the B of Q and all payments for fluctuations are based on them. The contractor should, if he considers the percentages incorrect, include the forecasted shortfall/overpayment in the adjustment item at the end of the B of Q.
2. Base index figures are ascertained by reference to indices applicable to the constituents of work 42 days prior to the date for the return of tenders.
3. Following the monthly valuation the current index figures are established taking those applicable 42 days prior to the date of the valuation. The 'price fluctuation factor' (PFF) is then calculated for each constituent of the work. All PFFs are added together to arrive at a single total for the contract for that particular month. The following formula is used to arrive at the PFF:

$$PFF = A \times \frac{(C - B)}{B}$$

    where A is the contract proportion, B the base index and C the current index.
4. Following the calculation of the PFF for the month the 'effective value' is calculated. This is the difference between:

    (a) the gross amount due to the contractor (before deducting retention) less any amounts for dayworks, nominated subcontractors, any item based on current cost and any previous fluctuations; and

    (b) the amount as calculated in (a) for the previous valuation.

5. When the effective value for the month has been calculated it is multiplied by the PFF for the month to arrive at the price fluctuation for the month.

Nominated subcontractors

Under clauses 58(1) and 58(2) the engineer has the power to order the contractor to employ a nominated subcontractor to supply goods, materials or services in respect of work covered by a prime cost or provisional sum in the B of Q.

Clause 59(3) states that the contractor is fully responsible for the work executed or goods supplied by nominated subcontractors (other than design). Contractors are thus not entitled to an extension of time for delay by a nominated subcontractor. However, clause 59(4) affords contractors some relief in permitting termination of the subcontract in the event of the nominated subcontractor's default.

There is no standard form of nominated subcontract under the ICE 6th, however the employer could use as a basis the domestic 'Blue Form' of subcontract published by the FCEC in September 1984.

The employer may wish to use the nomination procedure in order to choose who does the work, e.g. a plant specialist or structural steelwork supplier, or it may be appropriate where goods are on long delivery. However, because of the contractual pitfalls for the employer the nomination of subcontractors should be avoided if at all possible.

The main dangers to the employer in using the nomination procedures which could result in serious additional costs and delay include:

1. The contractor objects to the nomination. If the contractor exercises his right to object to a nomination then the employer must do one of the following:

    (a) nominate an alternative subcontractor – 59(2)(a);

    (b) vary the works under clause 51 – 59(2)(b);

    (c) omit the works and have it carried out under another direct contract either concurrently or at some other time – 59(2)(c);

    (d) arrange for the contractor to take over either by using a sub-contractor of his own choice – 59(2)(d) – or by executing the work himself – 59(2)(e).

2. Delay in nominating or renominating for which the employer is responsible.

3. Design by the nominated subcontractor which is not normally the responsibility of the contractor. A separate collateral warranty or direct agreement between the employer and nominated subcontractor is therefore required.

4. The risk of insolvency of a nominated subcontractor which is borne by the employer not by the contractor.

In accordance with clause 59(5) payment to the contractor for the nominated subcontractors should comprise:

1. The actual price due to be paid by the contractor to the subcontractor, except where due to the contractor's default, less all discounts except discounts for prompt payment (2.5 per cent on works contractors and 5 per cent on materials suppliers).
2. Any sum in the B of Q for labours in connection therewith.
3. The profit percentage inserted by the contractor in the B of Q.

Under clause 59(7) before issuing a payment certificate the engineer should be entitled to demand from the contractor reasonable proof that sums included in previous certificates in respect of nominated sub-contractor's work have been paid.

In the absence of proof, or good reason for not producing proof, the employer could pay the nominated subcontractor direct subject to certification by the engineer. The engineer could then deduct such direct payments from later payment certificates.

Settlement of disputes

If either the employer or the contractor have a dispute of any kind under the contract, including any decision of the engineer, they should serve a 'notice of dispute' to the engineer in accordance with clause 66(2). The engineer is then required to give his decision in writing to both parties within the time-scale specified.

Provided that the engineer has given his decision under clause 66(3) or the time for such decision has expired, and no notice to refer to arbitration has been made, either party may give notice in writing requiring the dispute to be considered under the ICE Conciliation Procedure (1988).

The conciliation procedure is intended to provide a means of settling disputes which is speedier and simpler to operate; the rules of the ICE Conciliation Procedure 1988 are included as a separate slip in the back of the ICE 6th.

If either party is dissatisfied with the decision of the conciliator the dispute can be referred to arbitration which should be conducted in accordance with the ICE Arbitration Procedure (1983).

This chapter, whilst not reviewing every single clause within the ICE 6th, has examined those subject areas which are likely to be of particular relevance to the client's contract administrator and the contractor's commercial manager.

Tutorial questions

1. Reference clause 2(1)(b); identify what matters could be included in item 18 in the Form of Tender (Appendix) on p. 49 of the ICE 6th
2. State the engineer's course of action following a contractor's persistent breach of clause 4(2)
3. Clause 44(1)(c) refers to 'any cause of delay referred to in these conditions'. Identify them

4. Clause 44(1)(e) refers to 'other special circumstances of any kind whatsoever which may occur'. Identify them

5. A contractor has entered a method-related charge in the B of Q prepared in accordance with the CESMM2 entitled 'Site concrete mixing plant – time-related charge; sum £25 000'.

   Explain how the engineer would decide the amount due to the contractor in monthly payments against this item:
   (a) if the contract proceeds as envisaged;
   (b) if there is a major variation to the works which increases the quantity of concrete to be placed by 25 per cent;
   (c) if the contractor decides to carry out the work using ready-mixed concrete and does not set up a concrete mixing plant on site

6. (a) Explain why an employer might wish to include nominated subcontractors.
   (b) Discuss the provision of the ICE Conditions of Contract (6th edition) in relation to nominated subcontractors with particular reference to the potential risks carried by the employer

7. Describe the procedures for payment to the contractor under the ICE Conditions of Contract and the redress available to the contractor if the engineer fails to certify or the employer fails to pay

8. Clause 11 of the ICE Conditions of Contract (6th edition) requires the contractor to inspect the site and obtain all necessary information regarding the sufficiency of his tender. Comment on the type of risks that may occur and the degree of comfort afforded the contractor by clause 11.

   You are the contractor's agent for a contract on which temporary sheet piles designed by your company are being driven. Details of a single borehole 100 m from the line of the piles was included with the tender documents and showed marl below the subsoil and a rate of boring which indicated to you that the material was reasonably soft. In fact the ground has proved too hard to drive the piles as required and an alternative and more expensive form of temporary works has to be adopted.

   Describe how you would seek to recover the additional costs thus incurred by your company

9. What are the contractor's obligations with respect to workmanship and materials under the ICE Conditions of Contract?

   Outline the powers given to the engineer in the ICE Conditions of Contract to ensure workmanship and materials comply with the contract

Questions 4–9 above are taken from the ICE Examinations in Civil Engineering Law and Contract Procedure. The examinations are held every June throughout the UK and overseas and can be taken by non-members of the ICE. Further details can be obtained from:
The Arbitration Office
The Institution of Civil Engineers
Great George Street
London SW1P 3AA
Telephone 071 222 7722

Further reading

Abrahamson M W 1979 *Engineering Law and the ICE Contracts Fourth Edition*, Elsevier Applied Science Publishers

Atkinson A V 1992 *Civil Engineering Contract Administration* 2nd edn, Hutchinson & Co. (Publishing) Ltd

Cottom G 1991 Better ICE conditions. *Construction News*, 7 February: 14

Cottom G 1991 ICE 6th keeps its balance . . . *Construction News* 14 February: 20

Eggleston B 1992 *The ICE Conditions of Contract: Sixth Edition – A User's Guide* Blackwell Scientific

Haswell C K, de Silva D S 1989 *Civil Engineering Contracts Practice and Procedure* 2nd edn. Butterworth

ICE Conditions of Contract 6th edition 1991

ICE Conditions of Contract 5th and 6th Editions compared Thomas Telford Ltd, 1991

Knowles R 1987 No obligation. *Chartered Quantity Surveyor* October: 7

Knowles R 1991 Engineering changes. *Chartered Quantity Surveyor* March: 16

Knowles R 1991 *The Institution of Civil Engineers Conditions of Contract 6th Edition: a Clause by Clause Guide* Knowles publications

Midlands Study Centre 1988 ICE Conditions of Contract, one-day course, 13 April, Albany Hotel, Birmingham

Mugurian G H 1991 6th Edition of the ICE Conditions – greater clarity means less likelihood of dispute but inconsistencies remain. *Civil Engineering Surveyor* March: 8–10

Mugurian G H 1992 *Guide to the Sixth, Explanatory Notes for Guidance to the ICE Conditions of Contract 6th Edition* ICES

Powell Smith V 1991 Improvements to Edition 6. *Contract Journal* 13 June: 9

Uff J 1993 *Construction Law* 5th edn Sweet & Maxwell

Webster M 1991 New Engineering Contracts – the 1991 Mancunian Lecture. *Civil Engineering Surveyor* September: 34–7

15

JCT Standard Form of Management Contract (1987 edition)

Introduction

Management contracting is not new, indeed it has been around for some 20 years. However, one of the major weaknesses of the system until the introduction of the JCT Standard Form of Management Contract (1987 edition) (JCT 87)[1] was the absence of standard conditions of contract; those offered by contractors often contained a number of fundamental differences which made tender assessment difficult.

Management contracting can be a more cost-effective and less adversarial approach for major projects. Under the system the contractor, who does not carry out any actual work on site, is concerned with the management of the project allowing the architect and other members of the professional team to concentrate on design.

The sixth RICS survey which took a 'snap shot' of the contracts in use in the building sector in 1993,[2] found management contracting forms being used on some 6.17 per cent of the value of the contracts whilst reflecting less than 1 per cent of the total number of contracts considered in the survey; the mean value of each management contract considered was £5m.

Practice Note MC/1 'Management Contract Documentation 1987' states that suitable conditions for use of the JCT 87 management contract may include:

- The employer wishes the design to be carried out by an independent architect and design team.
- There is a need for an early start and/or completion.
- The project is complex and/or large, over say £5m. at 1990 prices.
- Where it is intended that design and construction will overlap.
- The project might require changing the employer's requirements during construction, especially where the scope of work is difficult to define or is incomplete.
- The employer whilst requiring early project completion, wants the maximum possible competition in respect of the price for the construction works.

Procedures

Firstly the employer needs to appoint a 'professional team' comprising an architect/contract administrator who is the team leader, an engineer, a quantity surveyor and other specialist design consultants as required. The team then develop the project brief and calculate the budget price.

Appendix B to the Practice Note MC/1 sets out a possible method of inviting management contractors to tender. They should be invited at an early date, not later than the 'outline proposal stage' in the RIBA plan of work and the professional team should:

1. Identify the list of management contractors;
2. Invite written reaction to a brief description of the project;
3. Select a limited number of management contractors and provide them with a questionnaire about the project, its design, programme and cost aims: invite a written submission from management contractors including their fee proposals for:
   (a) total service expressed as a percentage of total project cost;
   (b) service for pre-construction period costs if the project is aborted before start on site.
4. Interview by employer and professional team with those selected management contractors;
5. Select successful management contractor based on:
   (a) experience of management contracting;
   (b) understanding of the project;
   (c) quality and experience of site project manager who will be present at the interview;
   (d) fee.

Whilst documentation is used as a basis of comparability of tenders for management contractors, interviews and subjective evaluation necessarily play a large part in the selection process to an extent not usually present in conventional tendering procedures.

Selection of works contractors usually follows conventional tendering procedures using drawings, bills, specifications as appropriate and generally involve standard single-stage, two-stage or design and build methods.

JCT 87 is not a lump sum contract and the sums payable will consist of items of prime cost (payable in respect of works contractors' accounts and the management contractor's own on-site management staff and facilities) together with the management contractor's fee.

In order to give an indication of the price the employer will pay (excluding the management contractor's fee) a 'contract cost plan' is prepared by the quantity surveyor based on the project drawings and specification the total of which is subject to consent by the management contractor. The contract cost plan once signed by the parties must be annexed to the contract (article 6.2).

The total of the contract cost plan and the management contractor's fee based thereon have to be agreed before any building work for the project can commence. It is the duty of the quantity surveyor and the professional team together with the management contractor to monitor regularly the 'prime cost' against the contract cost plan.

When the architect decides that the preparatory work in the pre-construction period has reached the stage when it will be practicable to commence the construction of the project he should notify the employer. The employer must then decide, within 14 days, either to notify the management contractor to proceed as a member of the management team or to notify him that he is not to proceed. This provision, called the 'break clause', enables the employer to determine the employment of the management contractor for any reason, e.g. because the project is no longer financially viable.

Duties of the management contractor

During the pre-construction period the management contractor will:

- prepare a detailed project and construction programme;
- prepare material flows and identify those components which require ordering early;
- advise on the practical implications of the proposed drawings;
- formulate and agree construction methods;
- advise on the layout and provision of site facilities and services;
- advise on the breakdown of the project into work packages;
- prepare lists of potential works contractors;
- assist in preparing documentation and obtaining tenders from works contractors.

The management contractor will not carry out any actual construction work on site; all site work will be executed by the works contractors. The management contractor's principal duties during the construction period are set out in the third schedule of the contract and include:

- planning and programming;
- cost estimating;
- monitoring the performance of the works contractors off and on site;
- cost control and payments;
- labour relations and quality control.

The management contractor should further provide the site facilities set out in the fifth schedule including the provision and maintenance of site offices and welfare facilities, hoardings, access roads, cranes, hoists and scaffolding, temporary services, health and safety and site cleaning, etc.

Essentially the management contractor will be required to manage, organise, supervise and secure the carrying out and completion of the project on or before the completion date. Payment is made by monthly interim certificates indicating the amount included for the works contractors and the amount of the construction period fee.

The forms

Standard Form of Management Contract 1987 edition
Conditions of contract between the employer and the management contractor for the construction of a building project.

Standard Works Contract

Conditions of contract between management contractor (M/C) and works contractors (W/C) comprising:

1. Works Contract/1 1987 Edition:
   (a) Invitation to tender (from M/C to W/C);
   (b) Tender by works contractor (from W/C to M/C);
   (c) Articles of agreement (which incorporate Works Contract/2 via article 1.2 of Part 3).
2. Works Contract/2 1987 Edition: works contract conditions (between M/C and W/C).
3. Works Contract/3 1987 Edition: standard form of employer/works contractor agreement.

JCT Subcontract/Works Contract Formula Rules 1987 Edition

In addition the Joint Contracts Tribunal have issued Practice Note MC/1 and MC/2.

Standard Form of Management Contract 1987

The Recitals and Articles of Agreement are followed by The Conditions of Contract, which comprise nine sections.

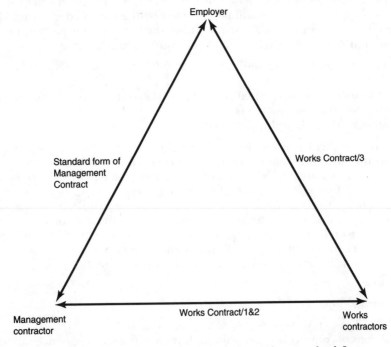

Figure 15.1 JCT Management Contract 1987 – standard forms

Section 1: Intention of the parties

This section includes definitions, statement of general and specific obligations of the management contractor, requirements relating to architect's instructions, contract documents and the issue and effects of certificates.

The contract provides for separate pre-construction and construction periods with the management contractor's fee expressed in two parts in the appendix.

The definition of works contract variation is similar to variation as defined in JCT 80; however, JCT 87 includes an additional provision for a project change which involves the alteration or modification of the scope of the project as shown on the project drawings and described in the specification.

In the event of late completion, provided there is no extension of time awarded, then theoretically the management contractor would be liable to liquidated damages. However, in accordance with clause 3.21 the management contractor would claim against the defaulting works contractor. However, such a deduction would be subject to the adjudication provisions of the works contract and the management contractor could not make instant deduction of any such sums.

In effect so long as the management contractor is not himself in breach and his employer remains solvent, he is guaranteed payment of all sums that he is liable for under the works contract. His only risk is that he may have to carry a temporary financial burden.

Section 2: Possession and completion

At the planning stage of the project the employer is required to give notice to the management contractor for him to proceed to the construction phase. The management contractor should indicate his consent to such a proposal by signing the contract documents whereupon the architect may give notice to the employer (with a copy to the management contractor) of the 'date when it will be practicable to commence the construction of the project'.

This section also contains an extension of time procedure similar to the JCT 63 provision under which the management contractor is required to notify the architect immediately it becomes apparent that the project completion is likely to be delayed. The architect's duty is to award a fair and reasonable extension of time for any of the identified events provided that the management contractor has used his best endeavours to reduce the delay. The grounds for extensions of time are:

- default of the employer or those for whom he is responsible;
- late issue of specifications, B of Q, instructions, drawings or levels for which the management contractor specifically applied in writing;
- relevant events under the works contract (similar to NSC/4 where the subcontractor is nominated under JCT 80 – pre-amendment No. 10).

Section 3: Control of the project

The architect's consent is required to the number and names of the contractor's management personnel identified prior to the construction period and the cost of such personnel will be included in the prime cost paid by the employer.

The architect's powers are extremely wide and far-reaching in that he can issue 'such instructions as are reasonably necessary to enable the management contractor to properly discharge his obligations'. In connection with variations the architect can instruct alterations to the scope of the project (project changes) and under the works contracts (works contract variations). The architect is also given powers to accelerate the progress of the project or alter its sequence or timing providing the employer has stated in the appendix that he wishes such a provision to apply.

Within this section are also contained important provisions setting out the rights and obligations of the management contractor and the employer when breaches of the works contracts in connection with design, materials or workmanship occur.

The management contractor is obliged to settle claims from works contractors, if necessary deducting the amount from sums due to other works contractors if it was their default which gave rise to the claim. If for any reason the management contractor fails to recoup the loss it should be made good by the employer provided that it was a loss for which the employer would be legally liable.

Section 4: Payment

During the construction period interim payment certificates, which are based on the quantity surveyor's valuation and paid on a monthly basis, would normally include the following items:

- amounts payable under works contracts after deducting 3 per cent retention;
- amounts for site staff and facilities and materials provided by the management contractor less 3 per cent retention;
- appropriate instalment of the construction period management fee, less 3 per cent retention;
- additional expenditure incurred by the management contractor due to breach of contract by the works contractor.

If any item of cost was 'incurred as a result of any negligence by the management contractor in discharging his obligations under the contract' and the negligence can be clearly established then the item would not be included in the prime cost.

The management contractor is required to submit, within 6 months of practical completion, all documentation to the quantity surveyor to enable him to ascertain the value of the prime cost. The quantity surveyor then has a further 3 months from receipt to deliver to the architect a statement of the prime cost and management fee.

Adjustment may be made to the construction period management fee if the prime cost exceeds or is less than the contract cost plan total by more than 5 per cent.

Section 5: Statutory obligations

Similar to JCT 80.

Section 6: Injury, damage and insurance

The provisions for the insurance of the project are modelled on the insurance and indemnity provisions of JCT 80 following amendment No. 2.

The management contractor is required, prior to commencement of any work on site, to take out a joint names policy (employer and management contractor) for all risk insurance cover for the project; this requirement applies whether the project is a new building or an alteration to an existing structure.

In the case of alterations to an existing structure the employer is required to take out a joint names insurance policy for existing structures and their contents.

Provisions are also included for the employer's loss of liquidated damages due to the occurrence of loss or damage due to a specified peril.

Furthermore, the management contractor is required to indemnify the employer in respect of personal injury or death or damage to property. The management contractor is required to take out, and ensure that each works contractor takes out, public liability insurance cover to the minimum sums as defined in the appendix.

Further provisions are also included requiring the management contractor to take insurance, if so instructed by the architect, for injury or damage to any property other than the works.

Section 7: Determination

This section again is not unlike the JCT 80 provisions. Four separate provisions are identified as grounds for the determination of the management contractor's employment:

1. Employer determines following defaults by the management contractor or the insolvency of the latter.
2. Management contractor determines following defaults by the employer or because of suspension of the project or the employer's insolvency.
3. Where either party determine following the suspension of the project due to a 'neutral event', i.e. not due to any default of either party.
4. Determination at will by the employer under which he is not required to give reasons.

Section 8: Works contractors

This section sets out the ground rules for the selection of the works contractors. The works contractors, who are selected following agreement between the architect and the management contractor, must be prepared to enter into the JCT Form of Works Contract unless the parties agree otherwise.

Provision is provided for early final payment to the works contractors provided that they have satisfactorily indemnified the management contractor against latent defects.

Section 9: Settlement of disputes – arbitration

The arbitration provisions are similar to those provided in JCT 80, amendment 4.

Additional sections of JCT 87 are as follows:

- an appendix divided into parts 1 and 2;
- VAT agreement;
- first schedule: description of project;
- second schedule: definition of prime cost payable to the management contractor;
- third schedule: services to be provided by the management contractor;
- fourth schedule; list of project drawings;
- fifth schedule: site facilities and services to be provided by the management contractor.

Works contracts

The works contract is split into three sections.

Works Contract/1 1987 Edition

This document consists of three parts:

Section 1: *Invitation to tender* which is completed by the management contractor and submitted to the works contractor together with the tender documents, i.e. drawings, specifications, B of Q, schedules of rates and full details of the contract including any changes to the standard form. An indication should also be given as to when the works are expected to be carried out on site.

Section 2: *Tender by works contractor* in which he provides details to the management contractor of tender sum, daywork percentages and formula rules. The tender, which may be on a lump sum or schedule of rates basis, should also state the periods required not only for off-site and on-site work but also for the submission of all necessary drawings.

Section 3: *Articles of agreement* which are signed or sealed by the management contractor and works contractor upon agreement. All the 'numbered documents' should be listed annexed together with the invitation and completed tender. The standard agreement notes that the Works Contract/2 conditions are deemed to be incorporated.

Works Contract/2 1987 Edition – works contract conditions

The Works Contract/2 conditions follow the format of the Standard Form of Management Contract being divided into nine sections; the only difference being an additional provision for fluctuations and section 8 in the works contract which deals with nominated suppliers. Many of the sections reproduce the equivalent provisions in the standard NSC/4 under JCT 80 – pre-amendment No. 10.

Works Contract/3 1987 Edition (Standard Form of Employer/Works Contractor Agreement)

This optional document resembles the standard NSC/2 Agreement between the employer and a nominated subcontractor under JCT 80 – pre-amendment No. 10 but is more restricted in its application.

Under Works/3 the works contractor warrants that he has exercised and will exercise all reasonable skill and care in the design of the works, selection of materials and the satisfaction of any performance specification, provided these matters are his responsibility. The works contractor also agrees to provide information to the architect in good time in order not to delay the project.

However, Works Contract/3 does not provide provisions for early payment to the works contract or for his early design work or payment for the advanced ordering of materials as the document is executed after Works Contract/1; if these facilities are required they should be the subject of a separate agreement between the employer and the works contractor.

Tutorial questions

1. State the powers and duties of the following parties under the JCT Standard Form of Management Contract:
   (a) employer
   (b) architect
   (c) management contractor
2. Discuss the extent of risk carried by the employer, management contractor and works contractors in connection with:
   (a) completion on time
   (b) total costs
   (c) quality of the project
3. Consider the employer and management contractor's liabilities in respect of a defaulting works contractor who becomes insolvent.
4. Discuss the premise that management contractors may be compromised in their relationships with professional advisers and works contractors

where they have general contracting origins. (RICS Final Examination, specimen paper (1987 syllabus), Quantity Surveying)

5. The management contractor is in delay under the management contract as the works contractor has been delayed by weather conditions. The works contractor is also in delay and has sought from the management contractor a 3 week extension of time because of the extremely wet conditions that prevailed. The management contractor has referred the matter to the contract administrator under the management contract.

The contract administrator has refused to grant an extension of time because he argues that had the management contractor complied with their own agreed programme no delay would have occurred. Notwithstanding this the amount of rainfall experienced was no more than should have been anticipated. The management contractor has consequently declined to grant an extension of time to the works contractor.

Discuss the position of the works contractor and the validity of the arguments for refusing an extension of time to both the management contractor and works contractor (RICS Final Examination 1990 (1987 syllabus), Quantity Surveying)

References

1. Joint Contracts Tribunal 1990 *Standard Form of Management Contract 1987 Edition Incorporating Amendments 1 and 2* Building Employers Confederation
2. Davis, Langdon, Everest 1994 *Contracts in Use: A Survey of Building Contracts in Use during 1993* RICS

Further reading

Austin S 1990 Calculating the risks. *New Builder* 8 March: 20, 21

Blacker T, Chappell D 1988 JCT management contract: an architect's guide (7-part article). *Architects' Journal* June/July. 1. The Agreement, 1 June: 68–72; 2. Conditions: Sections 1 and 2, 8 June: 71–6; 3. Conditions: Section 3, 15 June: 67–72; 4. Conditions: Section 4, 22 June: 61–4; 5. Conditions Sections 5 to 9, 29 June: 77–9; 6. Works Contracts 1 and 2, 6 July: 65–71; 7. Works Contracts/3, 13 July: 59–65

Hughes W P 1991 *An Analysis of the JCT Management Contract.* CIOB Technical Information Service, paper No. 130

Knowles R 1988 The new management contract (3-part article). *Chartered Quantity Surveyor* March: 11; April: 9; May: 7

Knowles R 1991 Terms of managing. *Chartered Quantity Surveyor* July: 8

Knowles R 1990 *The JCT Management Contract 1987: a Clause by Clause Guide to the Management Contract and Works Contract* Knowles Publications

NJCC *Code of Procedure for the Selection of a Management Contractor and Works Contractors* NJCC Publications, July 1991

16

FIDIC Fourth Edition (1987)

Introduction

The FIDIC (Fédération Internationale des Ingénieurs-Conseils) *Conditions of Contract for Works of Civil Engineering Construction*, Fourth Edition 1987, reprinted 1988 with editorial amendments (FIDIC 4th), is a truly international document.

The document is published by FIDIC whose headquarters is in Lausanne, Switzerland. Many suggestions and comments were received during the preparation of the Fourth Edition including those made by FIDIC members, international contractors, the European International Contractors, the Associated General Contractors of America, the World Bank, the Inter-American Development Bank, the Asian Development Bank and a group comprising Arab fund agencies.

However, the foreword to the FIDIC 4th states that whilst the conditions are intended for construction works invited on an international basis they can, subject to minor amendments, also be suitable for domestic contracts.

The Conditions of Contract, which are known as the 'red book', come in two parts:

Part I – General conditions, with forms of tender and agreement;
Part II – Conditions of particular application with guidelines for preparation of Part II clauses.

In the preparation of the conditions it was acknowledged that whilst there are many general clauses which would be applicable to all contracts many of the clauses would vary depending on the circumstances and locality of the works.

Part I is essentially a standard form in its entirety and is supplemented by the applicable clauses from Part II which is issued only as a guideline. The conditions of particular application should therefore be prepared individually for each contract and should be based on the Part II typical clauses or specially drafted for each particular project.

Part I of the Conditions of Contract at first sight follows the form and content of the ICE Conditions of Contract. It is a traditional contract intended for works of civil engineering construction where the permanent work is designed and supervised by a consultant engineer, though not necessarily the same consultant, and are measured and evaluated by B of Q. The FIDIC 4th is not suitable for lump sum or target contracts without some alteration and should not be used for entire design and build projects.

Those familiar with the ICE Conditions should not be misled into thinking that the FIDIC Conditions are a replication of the former. For example, FIDIC 4th is a form of contract not necessarily subject to the laws of England and some of the wording, definitions and procedures are different from those in the ICE Conditions.

A further major difference between the two forms is the prominence of the employer under the FIDIC form in the day-to-day activities on site particularly where financing the contract works is involved.

On major international projects it is not unusual to find the employer retaining control over the financial administration of the contract. This procedure is regularised in the FIDIC 4th where the employer is required to be consulted, along with the contractor, by the engineer under the clauses identified in Table 16.1, most of which are concerned with payment to the contractor and many with extensions of time.

Table 16.1   FIDIC 4th clauses requiring the employer to be consulted, along with the contractor, by the engineer

6.4	Delays and cost of delay of drawings
12.2	Adverse physical obstructions or conditions
27.1	Delay and costs of removing fossils, etc.
30.3	Transport of materials or plant – damage to bridge or road
30.4	Waterborne traffic – damage to structures
36.5	Engineer's determination where tests not provided for
37.4	Rejection (repetition of tests)
38.2	Uncovering and making openings
39.2	Default of contractor (removal of improper work)
40.2	Engineer's determination following suspension
42.2	Failure to give possession
44.1	Extension of time for completion
46.1	Rate of progress
49.4	Contractor's failure to carry out instructions (remedying defects)
50.1	Contractor to search
52.1	Valuation of variations
52.2	Power of engineer to fix rates
52.3	Variations exceeding 15 per cent
53.5	Payment of claims
64.1	Urgent remedial work
65.5	Increased costs arising from special risks
65.8	Payment if contract terminated
69.4	Contractor's entitlement to suspend work
70.2	Subsequent legislation

Further differences between the FIDIC 4th and the ICE 6th

The FIDIC 4th contains a cross-referenced index at the back of the document together with the editorial amendments which are summarised on the inside of the back flap. In the international form examples of the 'performance guarantee' and the 'surety bond for performance' are included in Part II of the conditions with the recommendation that where a performance security is required the form should be annexed to the conditions.

No inserts are included in the back flap of the FIDIC 4th covering the ICE conciliation procedure, the ICE arbitration procedure or a contract price fluctuation clause as in the ICE 6th.

The clause headings in the FIDIC 4th generally follow those in the ICE 6th; however the international form has additional clauses, detailed below.

Clause 34: Labour

The contractor is fully responsible for the engagement of all staff and labour including payment, housing, feeding and transport.

Clause 69: Default of the employer

The contractor is entitled to terminate his employment under the contract should the employer default in payment, interfere with a certificate, become bankrupt or find it impossible to continue due to economic circumstances. Upon such termination the contractor should remove all his equipment and be entitled to reimbursement of all reasonable costs.

If the employer fails to pay the contractor within the time specified the contractor might, after issuing the appropriate notice, choose to suspend or reduce the rate of work. If as a result the contractor suffers delay or incurred costs he would then be entitled to an extension of time and reimbursement of appropriate costs.

In contrast with the provisions in the FIDIC 4th it is noted that the ICE 6th provides only limited determination provisions for the contractor – when due to the employer's suspension of the works lasting more than 3 months.

Clauses 71 and 72: Currency and rates of exchange

If the government of the country in which the works are executed imposes currency restrictions any additional costs should be reimbursed to the contractor. However, the contractor carries the risk of changes to the rates of exchange if the contract provides for payments to be made in foreign currencies.

The exchange rates, determined by the central bank of the country in which the project is carried out, on a base date 28 days prior to the latest date for the submission of tenders are used if the tender is required to be expressed in a single currency and payments made in more than one currency.

Definitions and interpretations

Section 1 of Part I of the FIDIC 4th contains a list of definitions, many of which are essentially the same as those contained in the ICE 6th. Differences include:

1. 'Contract' means Parts I and II of the conditions, the specification, the drawings, the B of Q, the tender, the letter of acceptance, the contract agreement (if completed) and such further documents as may be expressly incorporated in the letter of acceptance or contract of agreement (if completed). In practice this may include pre-contract 'questions and answers' in which any ambiguities or anomalies in the lowest contractor's tender are identified and any price adjustment agreed between the parties.
2. A 'taking-over certificate' is issued by the engineer when the whole of the works have been substantially completed in accordance with clause 48; the wording used in the ICE 6th is 'certificate of substantial completion'.
3. 'Contract price' is defined as the sum stated in the letter of acceptance; in contrast the ICE 6th refers to the terms 'tender total' as the pre-contract price and 'contract price' as the final account price.
4. 'Plant' is defined as machinery, apparatus and the like intended to form or forming part of the permanent works. Contractor's plant is now referred to as 'contractor's equipment' in both the FIDIC 4th and the ICE 6th.
5. 'Cost' is defined as all expenditure properly incurred or to be incurred, whether on or off the site, including overhead and other charges properly allocatable thereto but does not include any allowance for profit; no mention is made of 'finance' as in the ICE 6th.

Engineer and engineer's representative

The FIDIC 4th contains no 'named individual' clause requiring the engineer to name the chartered engineer who will act as the 'engineer' under the contract as required under the ICE 6th.

Clause 2.3 of the FIDIC 4th states: 'The engineer may from time to time delegate to the engineer's representative any of the duties and authorities vested in the engineer.' Under the ICE 6th the engineer's representative is required to watch and supervise the construction and completion of the works, he has no authority to order any work involving delay or any extra payment or to issue variations and certain specified clauses cannot be delegated to him. There is no such limitation on the engineer's representative under the international form. However, many of the engineer's representative's decisions, as we have already seen, will be subject to consultation with the employer.

On major international projects the job titles of the client's PM team might cause confusion, e.g. on the HKMTR Stage 3 (Island Line) the 'construction manager' was delegated most of the powers of the 'engineer' and was responsible for a section of the works comprising four major projects. His immediate assistants, who would be responsible for two projects each, were referred to as 'senior resident engineers' (engineer's representatives). There were also 'resident engineers' (assistants) on each site often with an assistant (assistant's assistant!).

Clause 2.4 of FIDIC requires that the contractor is notified of the names, duties and scope of authority of such persons.

Clause 2.5 introduces a specific time limit of 7 days for confirmation by the contractor of any oral instructions made in writing by the engineer. If the engineer does not contradict such confirmation in writing within 7 days such instruction is deemed an engineer's instruction. Under a similar clause in the ICE 6th (clause 2(6)(b)) the terms 'as soon as is possible' and 'forthwith' are used.

Assignment and subcontracting

Under clause 4.1 of the FIDIC 4th the contractor is denied the right to subcontract any part of the works unless he has obtained the prior approval of the engineer. In contrast the ICE 6th merely requires that the contractor 'notify the engineer prior to the subcontractor's entry onto the site'.

An additional clause is provided in the FIDIC 4th covering the transfer to the employer of any guarantee provided to the contractor by a plant or services subcontractor valid after the end of the defects liability period (clause 4.2).

Contract documents

Part II of the conditions should state:

- the language or languages in which the contract documents are drawn up, and
- the country or state the law of which should apply to the contract.

Any ambiguities in the documents should be resolved by the engineer using the following priority:

1. The contract agreement (if completed)
2. The letter of acceptance
3. The tender
4. Part II of the conditions
5. Part I of the conditions
6. Any other documents forming part of the contract

However, if no order of precedence of documents is preferred Part II of the conditions suggests that subclause 5.2 could be varied to make the documents 'mutually explanatory of one another', i.e. as stated in clause 5 of the ICE 6th.

The courts have ruled that in the event of dispute the 'particular' should take precedence over the 'general'. If ambiguities are referred to the courts and cannot be resolved they will construe the contract against the party responsible for drawing it up, i.e. the employer (principle of *contra proferentem*).

General obligations

General responsibilities
The contractor is required to execute and complete the works and remedy any defects; he is also responsible for any design of the permanent works if so provided for in the contract; thus the provisions within the FIDIC 4th are similar to those in the ICE 6th.

Performance security
Where it is decided that a performance security is required clause 10.2 of the FIDIC 4th provides that such security should be valid until the contractor has executed and completed the works and remedied any defects in accordance with the contract.

Similarly under the ICE 6th the model form provides for release of the bond upon the issue of the defect correction certificate. However, in practice most contractors will offer an alternative bond which expires on the issue of substantial completion.

Inspection of site
In accordance with clause 11.1 the employer should make available to the contractor, before submission of the tender, such hydrological and subsurface data as have been obtained. The contractor is deemed to have examined the site before submitting his tender and taken into account the nature of the subsurface, the extent of the work, the means of access and the hydrological and climatic conditions. This latter item could be particularly relevant in overseas locations which may be prone to unpredictable typhoons, hurricanes, monsoons, blizzards, storms, etc.

Adverse physical obstructions or conditions
Clause 12.2 of the FIDIC 4th 'Adverse physical obstructions or conditions' is one of the most widely used clauses on major civil engineering projects: 'If, however, during the execution of the works the contractor encounters physical obstructions or physical conditions, other than climatic conditions on the site, which obstructions or conditions were, in his opinion, not foreseeable by an experienced contractor, the contractor shall forthwith give notice thereof to the engineer, with a copy to the employer.'

The clause in FIDIC refers to 'climatic conditions', as opposed to 'weather conditions' included in clause 12 of the ICE 6th. The FIDIC 4th requires the contractor to send a copy of the notice to the engineer and the employer.

Under the FIDIC 4th the contractor thus carries the risks arising from climatic conditions even though these may be difficult to establish in areas of unpredictable climate. However, clause 44.1 'Extension of time for completion', does provide some relief for exceptionally adverse climatic conditions in order to avoid reimbursement of liquidated damages.

It is further noted that under clause 12.2(b) of the FIDIC 4th the contractor is entitled to recover only the additional costs incurred by reason of the obstructions or conditions having being encountered; no mention is made of 'a reasonable percentage addition thereto in respect of profit' as clause 12(6) in the ICE 6th.

Programme to be submitted

Under clause 14.1 of the FIDIC 4th the contractor is required to submit a programme to the engineer for his consent, within the time stated in Part II of the conditions after the date of the letter of acceptance. The contractor is also required to submit, whenever required by the engineer, a general description of the arrangements and methods. In contrast, the ICE 6th requires the contractor to submit both his programme and method statement 'within 21 days after the award of the contract'.

Furthermore under the FIDIC 4th the form and detail of the programme should be prescribed by the engineer, whereas under the ICE 6th there is no mention of the form and detail required for the programme.

The ICE 6th requires that within 21 days of receipt of the programme the engineer either accepts it, rejects it stating reasons or requests the contractor to provide further information. In contrast, the FIDIC 4th is silent on the engineer's course of action after receipt of the contractor's programme or what he should do if he disagrees with it.

The author recollects on a major international tunnelling project the contractor submitted a programme based on grossly optimistic production rates bearing in mind the difficult ground conditions likely to be encountered. The contractor refused to submit a more realistic programme so it was left to the engineer's representative to prepare such a programme and it was this document which formed the basis for any evaluation of extensions of time.

Cash flow estimate to be submitted

Clause 14.3 of the FIDIC 4th requires the contractor, at the same time he provides the programme, to send to the engineer for his information a detailed cash flow estimate, in quarterly periods, of all payments to which the contractor will be entitled under the contract. These cash flows will be particularly useful for the project funding agents, e.g. an international financing institution. There is no requirement for cash flow estimates contained within the ICE 6th.

Safety, security and protection of the environment

Clause 19.1(c) of the FIDIC 4th requires the contractor to 'take all reasonable steps to protect the environment on and off site'. A similar provision in the ICE 6th is not so clearly defined.

Care of the works and insurance

Clause 20 ('Care of works') in the FIDIC 4th generally follows the format of the ICE 6th with the contractor required to take full responsibility for the care of the works, materials and plant. The contractor is not responsible for the excepted risks which are borne by the employer. These 'employer's risks' are generally as those defined in clause 20(2) of the ICE 6th with the following amendments:

- 'riot, commotion or disorder, unless solely restricted to employees of the contractor or of his subcontractors and arising from the conduct of the works' (FIDIC 4th, clause 20.4(e));

- 'any operation of the forces of nature against which an experienced contractor could not reasonably have been expected to take precautions' (FIDIC 4th, clause 20.4(h)).

In accordance with clause 21.1 in the FIDIC 4th the contractor is required to insure:

1. The works together with materials and plant;
2. An additional sum of 15 per cent of such replacement cost to cover for professional fees and demolition (the ICE 6th requires a 10 per cent premium);
3. The contractor's equipment (this requirement is not specifically expressed in the ICE 6th).

The insurance for parts (1) and (2) is required to be in the joint names of the contractor and employer and should cover both parties until the taking-over certificates and the contractor during the defects liability period.

Clause 22.1 of the FIDIC 4th requires the contractor to indemnify the employer against losses in respect of death or injury to persons or loss or damage to any property (other than the works). Again there are 'exceptions' which are not required to be indemnified by the contractor and are similar to clause 22(2) of the ICE 6th; the FIDIC 4th, however, makes no mention of 'damage to crops being on the site' as mentioned in the ICE 6th.

Clause 23 of the FIDIC 4th requires the contractor to take out the necessary 'third party insurance', in the joint names of the contractor and the employer, against liabilities for such death or injuries of any person or loss of or damage to any property (other than the works).

Clause 24 states that the employer is not responsible for any damages or compensation payable to any workmen in the employment of the contractor or his subcontractors; the contractor should therefore take out the necessary insurance, or ensure that his subcontractors do likewise.

In accordance with clause 25 the contractor is required to provide evidence that the insurances have been effected before the work commences on site; the full policy document must be produced within 84 days of the commencement date.

Part II of the conditions states that the employer might wish to arrange insurance of the works and third party insurance, for example where a number of separate contractors are employed on a single project or phased handover is involved.

Details of any insurance taken out by the employer should be included in the tender documents in order that the contractors can assess what additional insurance should be included in their bids.

Labour

The FIDIC 4th contains clause 34.1 'Engagement of staff and labour' requiring the contractor to make his own arrangements for the engagement of all staff, local or otherwise, and for their payment, housing, feeding and transport.

Part II of the FIDIC 4th advises that it will generally be necessary to add further subclauses under clause 34 depending on the circumstances and locality of the works. The examples of subclauses given in Part II give a real flavour of the nature of the challenge and environments which might be encountered on major international civil engineering projects:

1. Repatriation of labour – 'The contractor shall be responsible for the return to the place where they recruited . . .' e.g. labour recruited in developing countries for projects in the Middle East.
2. Measures against insects and pests – 'The contractor shall warn his staff and labour of the dangers of bilharzia and wild animals.'
3. Arms and ammunition – 'The contractor shall not give, barter or otherwise dispose of to any person or persons, any arms or ammunition of any kind . . .'
4. Festivals and religious customs – 'The contractor shall in all dealings with his staff and labour have due regard to all recognised festivals, days of rest and religious or other customs.'

Materials, plant and workmanship

The contractor is required to provide all materials, plant and workmanship in accordance with the contract and the engineer's instructions.

If any tests required by the engineer, additional to those provided for in the contract, show the materials, plant or workmanship not in accordance with the contract then the costs are borne by the contractor. However, clause 36.5 of the FIDIC 4th states that if such tests show the work in accordance with the contract clause then the contractor may be entitled to an extension of time together with any additional costs. The equivalent clause in the ICE 6th (clause 36(3)) contains no specific reference to extension of time entitlement.

Under the FIDIC 4th the engineer is required to give the contractor 24 hours' notice of his intention to carry out the inspection or to attend the tests. If the engineer determines that the materials, etc. are defective he should notify the contractor immediately stating the reasons; the contractor should thereupon make good such defects as soon as possible.

Clause 37.5 enables the engineer to delegate the inspection and testing to an independent inspector who will be considered an assistant of the engineer.

Suspension

In accordance with clause 40 of the FIDIC 4th if the progress of the works is suspended, on the instructions of the engineer, for more than 84 days and such suspension is not:

1. Provided for in the contract, or
2. Due to some default of the contractor, or
3. Due to climatic conditions on the site, or
4. Necessary for the proper execution of the works or the safety of the works

then, if permission to resume is not given by the engineer, the contractor could give notice to the engineer requiring permission to proceed with the works. If within 28 days of receipt such permission is not granted the contractor could elect to treat the section or whole of the works as an omission of the contract work under clause 51 or as a default by the employer and terminate his employment under the contract.

The ICE 6th has a similar provision for suspension in clause 40 but refers to a suspension lasting more than 3 months compared with the 84-day period under the FIDIC 4th.

Commencement and delays

The contractor should commence the works as soon as possible after the receipt of notice from the engineer; such notice should be issued within the time stated in the appendix after the date of the letter of acceptance.

Clause 41 in the ICE 6th states that the works commencement date is either specified in the appendix, or within 28 days of award of the contract, or agreed between the parties.

The grounds for extension of time for completion under the FIDIC 4th are specified as:

* extra or additional work
* any cause of delay referred to in the conditions
* exceptionally adverse climatic conditions
* any delay, impediment or prevention by the employer
* other special circumstances, other than the contractor's default

The ICE 6th specifically identifies increased quantities (not due to a variation) as grounds for an extension of time. It does not specifically mention prevention by the employer or exclude defaults of the contractor under 'other special circumstances'.

Under the FIDIC 4th the contractor should notify the engineer within 28 days, with a copy to the employer, that the event has first arisen and send full details to the engineer within a reasonable time as agreed by the engineer.

Alterations, additions and omissions

The engineer is given the power to instruct variations to the works similar to the provisions under the ICE 6th. However, in accordance with clause 51.1(b) in the FIDIC 4th the engineer is not empowered, without the contractor's agreement, to omit from the contract any part of the work if the reason for this is that the employer now wishes to carry out such work or wishes to employ another contractor to do so. Nor is the engineer empowered under the international form to instruct variations after the works have been substantially completed.

The provisions for the fixing of rates where the work is varied, or where due to a variation the rates for the other work is rendered inappropriate, are similar to those contained in the ICE 6th. In accordance with clause 52.1 of the FIDIC 4th, the engineer is required to 'determine provisional rates and prices to enable on-account payments to be included in certificates'; the intent in the ICE 6th is similar but it is not so clearly spelt out.

Clause 52(2) of the FIDIC 4th requires that either the contractor or the engineer notify the other if they intend to claim a varied rate due to a variation within 14 days of the date of such an instruction, before the commencement of the varied work. Again the procedure under the ICE 6th is not so strict – the contractor should preferably notify the engineer before the varied work is commenced or 'as soon thereafter as is reasonable in all the circumstances' (ICE 6th, clause 52(2)).

Where the value or variations or remeasurement on completion has varied the value of the works by more than 15 per cent a sum may be added or deducted from contract price to take account of the contractor's site or general overhead costs. Such a sum should be based only on the amount by which such additions or deductions are in excess of 15 per cent of the 'effective contract price' (tender price less provisional sums and allowance for dayworks). This clause (52.3) is peculiar to the FIDIC Conditions and finds no counterpart in the ICE 6th.

Clause 52.4 (daywork) enables the engineer to instruct that varied work to be valued on a daywork basis to be valued under the terms set out in the daywork schedule included in the contract. Unfortunately in practice this schedule can prove to be less than comprehensive; often key items of labour and construction equipment occur on daywork sheets with no specific items include in the schedule.

No mention is made in the FIDIC 4th of the *FCEC Schedule of Daywork Charges* which when used with *The Surveyor's Guide to Civil Engineering Plant* provides a comprehensive schedule of rates for construction equipment. However in practice these two documents can prove useful, on international works, when calculating rates for construction equipment.

Procedure for claims

At first sight the claims procedures under the FIDIC 4th appear practically the same as those required under clause 52(4) of the ICE 6th. However, upon second reading it is clear that the provisions under the FIDIC 4th require the contractor to keep to a more strict timetable as follows:

- Contractor to give notice to the engineer, with a copy to the employer, within 28 days of the event giving rise to the claim (53.1).
- Engineer upon receipt of notice can instruct contractor to keep any further records; contractor to provide records (53.2).
- Contractor to send to the engineer an account giving detailed particulars within 28 days, or such other reasonable time as may be agreed by the engineer (53.3). (ICE 6th refers to 'as soon as reasonable in all the circumstances'.)
- Contractor shall send further interim accounts giving accumulated amount of claim and further grounds, at intervals as the engineer may reasonably require (53.3).
- Contractor to send a final account within 28 days of the end of the

effects resulting from the event (53.3). (ICE 6th – no mention of time-scale for receipt of the final account of the claim in clause 52(4)).

- If contractor fails to comply with the procedure his entitlement to payment shall not exceed the amount verified by contemporary records.

Measurement

Clause 55.1 states that 'The quantities set out in the bill of quantities are the estimated quantities of the works', thus implying some sort of remeasurement contract. However, little guidance is given in clause 57.1 as to the method of measurement, merely that the works shall be measured net.

In contrast, the ICE 6th provides for a detailed description of the work in accordance with the Civil Engineering Standard Method of Measurement Second Edition 1985 (this could be amended to Third Edition or Highways Method of Measurement).

Furthermore, clause 57.1 in the FIDIC 4th requires the contractor to submit to the engineer, within 28 days after receipt of the letter of acceptance, a breakdown of the lump sum items contained in the tender. This breakdown is presumably required merely for the purpose of interim payments and not for the valuation of variations.

At first sight the above clauses may seem vague and contradictory. However, in practice the extent of remeasurement of the project may vary depending on the nature of the works as shown in Table 16.2.

Clause 56.1 of the FIDIC 4th requires the engineer to remeasure and determine the value of the works. Again the provisions require the contractor to adhere to a more rigorous timetable than that prescribed under clause 56(3) in the ICE 6th:

- Should the contractor fail to attend such measurement then the measurement taken by the engineer shall be taken to be the correct measurement.

Table 16.2 Examples of typical measurement provisions within FIDIC contracts

Case Study No. 1 Marine terminal in the UK, American employer
Activity schedule, prepared by the engineer, describing major elements of the project included in the tender documentation; contractors required to insert lump sum prices against activities

Piling and pile-anchoring activities required lump sums to be further analysed within a schedule of approximate quantities prepared by the engineer with rates inserted by the contractors against each item; subject to remeasurement on completion

Case Study No. 2 HKMTRC – Stage 3 (Island Line)
All work subject to remeasurement in accordance with the employer's own method of measurement (essentially the same as the CESMM2 with a modified tunnelling section and the addition of architectural work)

- The contractor has 14 days in which to agree to the records so produced, if he fails to attend to examine such records then they should be taken as correct.
- If after examination the contractor does not agree with such records he should notify the engineer within 14 days of such examination.

Certificates and payment

Clause 60.1 of the FIDIC 4th requires the contractor, after the end of each month, to submit to the engineer six copies of a statement showing the amounts to which the contractor considers himself entitled to payment. The statement should include:

1. The value of the permanent works executed;
2. Any other item in the B of Q, including contractor's equipment, temporary works, dayworks and the like;
3. The percentage of the invoice value of listed materials, all as stated in the appendix, and plant on the site but not incorporated in the works;
4. Adjustments under clause 70 (changes in cost and legislation);
5. Any other sum to which the contractor may be entitled under the contract.

No provision is made in the FIDIC 4th for payment for materials identified in the appendix which have not been delivered to site but are vested in the employer (ICE 6th, clause 60(1)(c)).

The international form requires retention, as stated in the appendix, to be applied to items 1–3 and 5 of clause 60.1. It would thus appear that retention is deducted twice in the case of materials on site.

Under the FIDIC 4th the engineer is required to certify payment within 28 days of receiving the statement (clause 60.2) and the employer must make payment to the contractor within 28 days after such interim certificate has been delivered to the employer.

Thus in theory the contractor could wait 56 days after the end of the month before receiving payment, which would obviously have serious repercussions on his cash flow. In contrast, the ICE 6th requires that within 28 days of the date of delivery to the engineer . . . of the contractor's monthly statement the engineer shall certify and the employer shall pay the contractor.

Not later than 84 days after the issue of the 'taking-over certificate' the contractor should submit to the engineer a 'statement at completion' stating the final amount of work done in accordance with the contract together with any further sums which the contractor considers due and an estimate of any sums which may become due.

Not later than 56 days after the issue of the 'defects liability certificate', the contractor must submit a 'draft final statement' to the engineer stating the value of all works done in accordance with the contract and any further sums which the contractor considers due to him. The engineer is then required to issue a final certificate within 28 days after receipt of the final statement.

The ICE 6th makes no mention of the requirement for a statement on completion; it does, however, require the contractor to submit to the engineer a statement of final account, not later than 3 months after the date of the defects correction certificate.

Clause 60.7 of the FIDIC 4th requires upon submission of the final statement the contractor gives to the employer, with a copy to the engineer, a written discharge confirming that the total of the final statement represents full and final settlement of all monies due to the contractor arising out of or in respect of the contract. No such discharge is contained within the ICE 6th.

Part II of the Conditions of Contract refers to the following additional subclauses which could be added to the contract to cover certain other matters relating to payment:

- where payments made in various currencies in predetermined proportions and calculated at fixed rates of exchange;
- where payments are to be made in one currency;
- where an advanced payment of an amount stated in the appendix to tender is required to be made to the contractor, i.e. a sum for mobilisation.

Settlement of disputes

Under clause 67.1 of the FIDIC 4th if a dispute of any kind arises between the employer and the contractor it should be referred in writing to the engineer with a copy to the other party; the engineer then has 84 days to make a decision.

If the employer or the contractor is dissatisfied with the decision of the engineer, or if the engineer fails to respond within the time stipulated, then within 70 days from such notice or the expiration of the time stipulated, an arbitration notice may be served by either party.

Clause 67.2 requires that the parties should attempt to settle the dispute amicably prior to the commencement of an arbitration. The arbitration should not commence until on or after the fifty-sixth day after the day on which the notice of intention to commence the arbitration was given, thus allowing a 'cooling-off' period for amicable settlement.

Any arbitration should be settled, unless otherwise specified in the contract, under the rules of conciliation and arbitration of the International Chamber of Commerce.

There is no provision in the international form for the intermediate stage of conciliation, prior to or in place of arbitration, as clause 66(5) of the ICE 6th.

Changes in cost and legislation

Clause 70.1 requires that any adjustments to the contract price for rise or fall in the cost of labour and/or materials should be determined in accordance with Part II.

Clause 70.2 enables the contractor to recover/reimburse any additional/

reduced costs, other than under clause 70.1, due to subsequent changes in legislation. Such legislation could be national, state, or by-laws and must occur after 28 days prior to the latest date for submission of tenders.

Part II provides for three alternative methods of dealing with price adjustment:

1. Suitable where a contract is of short duration: subject to clause 70.2 – contract price not subject to any adjustment; or
2. Price adjustment to be made for the difference in price between the basic price and the current price of local labour and specified materials (similar to the traditional fluctuations system as clause 39 of JCT 80);
3. Adjustments to the contract price for rise or fall in labour and materials to be made monthly based on published indices (similar to the contract price fluctuations provisions under the ICE 6th).

This chapter, whilst not reviewing every single clause within the FIDIC 4th, has examined those subject areas which are likely to be of particular relevance to the client's contract administrator and the contractor's commercial manager.

Tutorial questions

1. List the type of information which should be included by the contractor in a daily site diary identifying the relevant clauses in the FIDIC 4th
2. Identify the clauses in the FIDIC 4th which provide the contractor with opportunities for increasing the contract price; state whether extra costs, extra time and profit are applicable
3. Identify those clauses in the FIDIC 4th which entitle the employer to receive payment from the contractor; state whether an engineer's notice or no notice is required to be given
4. Part II of the FIDIC 4th contains the optional clause 47.4 'Bonus for (early) completion'; state:
   (a) What are the advantages and disadvantages of including the bonus clause in the contract?
   (b) Under what circumstances would you recommend/not recommend the use of this clause?
   (c) What is the basis for the calculation of the daily rate for the bonus?
   (d) What other measures (apart from bonus payments) could be used if completion on time was critical to the success of a multi-disciplinary major international project?
5. 'In accordance with the provisions of clause 52.3 the engineer, after consultation with the employer, would be entitled to reduce the contract prices if the effective contract price is increased by more than 15 per cent.'
   Comment on this statement

Further reading

Eggleston B 1993 *The ICE Conditions of Contract: Sixth Edition, a user's guide* Blackwell Scientific

FCEC 1990 *Schedules of Dayworks Carried out Incidental to Contract Work*

FIDIC *Conditions of Contract for Works of Civil Engineering Construction*, Part I General Conditions, Part II Conditions of Particular Application, Fourth Edition 1987, reprinted 1988 with editorial amendments

FIDIC 1990 *Guide to the Use of the FIDIC Conditions of Contract for Works of Civil Engineering Construction*

Frick-Meijer S E et al 1988 FIDIC conditions of contract for work of civil engineering construction, fourth edition. *Proceedings Institution of Civil Engineers* Part 1, **84,** August: 821–36.

ICES 1991 *The Surveyor's Guide to Civil Engineering Plant*

Knowles R 1988 FIDIC – new revision. *Chartered Quantity Surveyor* February: 6, 8

Rushbrooke P L 1983 *Working with FIDIC – a Practical Approach to its Use in the Middle East* CIOB

Sawyer J G, Gillott C A 1990 *The FIDIC Digest Contractual Relationships, Responsibilities and Claims under the Fourth Edition of FIDIC Conditions* Thomas Telford

17

New Engineering Contract

Introduction

In 1985 the Institution of Civil Engineers (ICE) decided to instigate a fundamental review of alternative contract strategies for civil engineering design and construction in order to identify good practice. They identified the following changes which have occurred in recent years:

- the construction industry had become far more adversarial which in turn had led to a diversion of talent;
- increased involvement of contractors in the design process particularly where time was of the essence;
- the introduction and success of management contracting in the building sector particularly on major projects;
- proliferation of different standard forms of contract in the process, M&E, civil engineering and building sectors;
- the continuing erosion of the engineer's 'independence' under traditional ICE Conditions of Contract;
- development of project management (PM) as a science and the gradual introduction of the project manager into the construction process;
- a steady decline in the influence of UK practice internationally;
- a better appreciation of risk allocation in different contract strategies.

Furthermore it was considered that with the introduction of the single European market in 1992 it was essential to consider the best continental practices.

The first objective of the New Engineering Contract Working Group was to produce a flexible form of engineering contract which could be used as the basis of a management contract, a construction management agreement or design and build, as well as for conventional lump sum contracting.

After several years of discussion and feedback from interested parties the New Engineering Contract (NEC) was introduced as a 'consultative document' in 1991. Later after considering feedback from users of the original form the First Edition of the contract was introduced in 1993.

The NEC shows a radical departure from traditional construction contracts; indeed learned opinion considers that it may well remove all the time-honoured concepts and guidelines which have evolved over the past 100 years.

A significant factor of the NEC is its length; it is claimed that a typical contract under the NEC will contain only 10 000 words, compared to 25 000 under the ICE 6th and 45 000 under JCT 80.

Much of the terminology in the NEC is new; the specification and drawings become the works information, extensions of time become compensation events and the maintenance period becomes the defects correction period.

Achievement of objectives through key features

The task of the drafting team for the NEC was to design a contract which was flexible, clear, simple and which stimulated good management of projects.

Flexibility

Flexibility is achieved by ensuring that the contract is suitable for many applications – civil, electrical, mechanical and building. It is intended to be used whether the contractor has some design responsibility, full design responsibility or no design responsibility.

It could be used on multidisciplinary projects, management, target or cost reimbursable or traditional consultant designed works incorporating an admeasurement contract. It is designed for use in the UK but could easily be adapted for work abroad.

Clarity and simplicity

Clarity is achieved in the NEC by the use of straightforward English, with few long words, shorter sentences and avoiding the use of such words as 'reasonable' and 'to the satisfaction of the engineer'.

Flow charts have also been produced as well as a set of guidance notes in order to aid the familiarity and help in the administration of the projects.

Assimilation of the clauses contained within the NEC have also been made easier by the order in which they are presented. Clauses of a like nature have been grouped together in nine sections identified as 'core clauses' comprising: general, the contractor's main responsibilities, time, testing and defects, payment, compensation events, title, risks and insurance, disputes and termination.

Two further changes from traditional contracts require specific mention. Firstly, the NEC does not contemplate the use of nominated subcontractors. However, the form has been designed so that an employer who has reasons for using a particular contractor for part of the works can use the NEC for a direct contract alongside other contractors. Subcontractors are accommodated under the NEC and a form of subcontract has been produced which is fully compatible with the main contract.

Secondly, the financial control document in the NEC can be either a traditional B of Q or an 'activity schedule'. The activity schedule is supported by a construction programme which includes a method and resource statement. There is no provision for a quantity surveyor within the NEC.

Sound project management

One of the most important characteristics in the NEC is the emphasis on sound PM in an attempt to avoid the adversarial nature of traditional contracts. The NEC is founded on the proposition that foresighted co-operative management between the parties can shrink the risk inherent in construction work and lead to a successful outcome.

An example of this collaborative approach is found in core clause 16 'Early warning' which requires either the contractor or the project manager to give the other an early warning as soon as either becomes aware of any matter which may increase the total of the prices, delay completion or affect the performance of the works in use.

The approach to the management of 'compensation events' which includes variations is also innovative. On a conventional contract variations are frequently valued using the rates in the B of Q which can often lead to dispute.

Under the NEC the contractor is required to consider alternative approaches and submit quotations for dealing with problems before the event. The contractor must include for the full effect of the variation including all possible delay and disruption costs in his quotations, the theory being that the contractor will then be indifferent to any choice made by the employer.

The project manager will then select the appropriate method after consultation with the employer. Once the quotation for the variation is accepted by the project manager, the contractor then retains the incentive to complete the work in an efficient manner and the risk to the employer is minimised.

This approach is intended to stimulate foresight and few issues relating to the valuation of work or extensions of time need be left until after the event.

Other 'compensation events', e.g. adverse physical conditions, late instructions, should be dealt with on the basis of the 'actual cost' of carrying out the work. What can and cannot be included is defined in the 'schedule of cost components' which forms part of the contract. To the actual cost is then added a 'fee percentage' which covers the contractor's off-site costs and profit; the fee percentage is part of the contractor's competitive tender for the contract.

The structure of the NEC

Standard construction forms of contract are generally inflexible, requiring the implementation of a fixed set of clauses irrespective of the nature of the project. This has resulted in an increasing tendency for standard contract

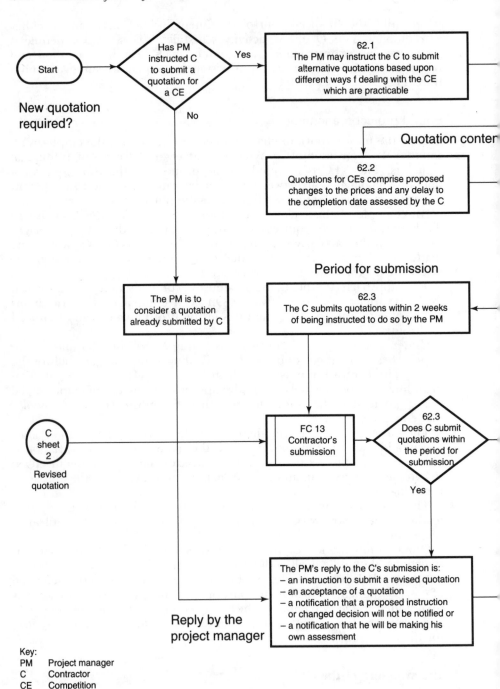

New quotation
required?

Quotation conter

Period for submission

Revised
quotation

Reply by the
project manager

Key:
PM   Project manager
C   Contractor
CE   Competition
FC 63 Flowchart for clause 63

Figure 17.1   NEC flowchart 62. Quotation for compensation events – sheet 1
of 2. PM = project manager; C = contractor; CE = compensation event. *Source:*
New Engineering Contract. Reproduced by permisson of the Institution of
Civil Engineers

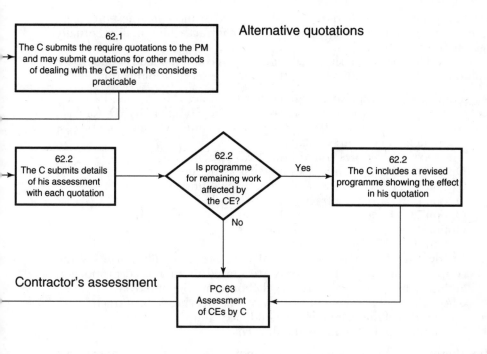

Alternative quotations

62.1
The C submits the require quotations to the PM and may submit quotations for other methods of dealing with the CE which he considers practicable

62.2
The C submits details of his assessment with each quotation

62.2
Is programme for remaining work affected by the CE?

Yes

62.2
The C includes a revised programme showing the effect in his quotation

No

Contractor's assessment

PC 63
Assessment of CEs by C

No

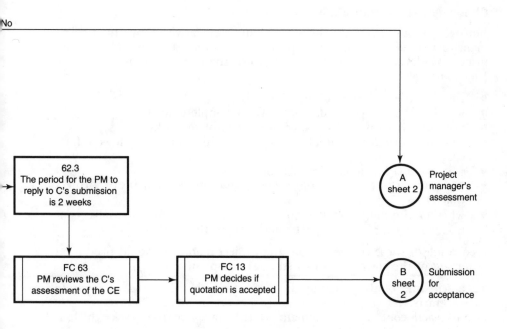

62.3
The period for the PM to reply to C's submission is 2 weeks

A
sheet 2

Project manager's assessment

FC 63
PM reviews the C's assessment of the CE

FC 13
PM decides if quotation is accepted

B
sheet 2

Submission for acceptance

clauses to be amended to suit the client and project in question. Such alterations can create an imbalance in the contract which may eventually lead to a dispute situation.

The NEC contains ten slim unnumbered but colour coded A4 paperback volumes – costing £10.00 each or the complete set in a folder for £55.00.

- Volume one – the New Engineering Contract – is intended for use as a reference document; it contains the complete NEC contract core clauses, optional clauses (main and secondary), schedule of cost components and contract data.
- The next six volumes contain the NEC options A to F:

  Option A   Priced contract with activity schedule
  Option B   Priced contract with bill of quantities
  Option C   Target contract with activity schedule
  Option D   Target contract with bill of quantities
  Option E   Cost reimbursable contract
  Option F   Management contract

  (In options A and B the financial risks are generally borne by the contractor, in C and D the risks are shared in an agreed proportion and in E and F the risks are largely borne by the employer.)
- The eighth volume contains the New Engineering Subcontract.
- The ninth volume contains the Guidance Notes.
- The tenth volume contains the flow charts.

It is therefore apparent that the NEC is more a family of different contract forms rather than one contract form as the name might suggest.

Choosing the appropriate option

Before any design commences the employer (advised by the project manager) should choose the appropriate contract strategy. The factors to be considered are identified on p. 13 of the brown volume 'Guidance Notes', namely:

- who has the necessary design expertise;
- whether there is particular pressure to complete quickly;
- how important is performance of the complete works;
- whether certainty of final cost is more important than lowest final cost;
- what total risk is tolerable for contractors;
- how important cross-project co-ordination is to achieve project objectives;
- whether the employer has good reasons for selecting specialist contractors or suppliers for particular parts of the work.

As a result of these considerations the project manager should produce a statement of the chosen contract strategy comprising:

- a schedule of the parts of the project which will be let as separate contracts;
- for each contract, a statement of the stages of the work which it

will include, covering management, design, manufacture, erection, construction, installation, testing and commissioning as appropriate;

- a statement of the NEC main option (A to F) to be used for each contract.

All the main options can be used with the boundary between design by the employer and design by the contractor set to suit the chosen strategy.

Further choices to be made

Having selected the contract option (A to F), it will then be necessary for the user to select the appropriate secondary options (the optional clauses). It is not necessary to use any of them and any combination, other than those excluded, may be used.

Option G   Performance bond
Option H   Parent company guarantee
Option J   Advanced payment to the contractor
Option K   Multiple currencies (not to be used with Options C, D, E and F)
Option L   Sectional completion
Option M   Limitation of the contractor's liability for design to reasonable skill and care
Option N   Price adjustment for inflation (not be used with Options E and F)
Option P   Retention (not to be used with Option F)
Option Q   Bonus for early completion
Option R   Delay damages
Option S   Low performance damages
Option T   Changes in the law
Option U   Special conditions

Each of the six contract option volumes (A to F) contain a 'cocktail' of clauses comprising:

- core clauses
- main option clauses (these are included in sequence after the core clauses and are printed in bold type)
- secondary option clauses

Additional information in each volume (A to F) comprises:

- two schedules of cost components (except in Option F – the management contract);
- contract data, a nine-page critical document containing a nine-page menu of items, viz:
  part one – data provided by the employer: name and address of parties, contractor's liability for design, date for possession, programme requirements, testing and defects, payment terms including interest rate for late payments, weather data, insurance requirements, name of adjudicator, etc.

part two – data provided by the contractor: list of key people, tendered total, works information for contractor's design, option for contractor to state completion date, hourly rates, etc.

By structuring the contract in this way there is less reason to alter the form. Provision is made for the inclusion of other clauses to suit the client under Option U 'Special conditions'. Guidance is provided as to which clauses are compatible with each other.

The contract documents making up the NEC comprise:

- form of tender
- the schedule of contract data
- the core clauses
- the optional clauses ⎫ The conditions of contract
- the schedule of actual cost

The role of the engineer

Several shortcomings in the traditional role of the engineer under the ICE Conditions of Contract have been identified. Many employers had conceded 'the engineer' less independence than required under the contract. This may have been achieved through the influence of auditors or the constraints placed on the engineer when he is a direct employee of the employer. There was also the engineer's dual role as manager of the project on behalf of the employer and adjudicator of disputes which he himself may have caused. Furthermore, it was apparent when using the ICE Conditions of Contract that the employer had very little influence over the day-to-day running of the project.

The traditional role of the engineer is divided in the NEC into its four constituent roles:

1. Project manager
2. Designer
3. Supervisor of construction
4. Adjudicator of disputes

The role of the project manager is to manage the project on behalf of the employer. The contract empowers the project manager to instruct the contractor, certify payments and completion, approve programmes and contractor's designs and even to accelerate the works.

The project manager is constrained from acting unreasonably by statements in the contract. If the contractor believes that any of the project manager's instructions are not in accordance with the contract he may refer them to the adjudicator.

The principal role of the supervisor is to monitor the contractor's performance for compliance with the specification and drawings, called collectively the works information.

The adjudicator, who is named in the contract and is independent of the employer, carries out the function formerly carried out in many contracts by the engineer in his quasi-arbitral role.

Balance of risk

The balance of risk carried by contractors and employers in the NEC is broadly speaking similar to that within the standard ICE Conditions. Interestingly, the NEC core clause 80 'Employer's and contractor's risks' identifies those risks that are allocated to the employer.

Additionally all the items which may give rise to a request for additional time or costs are identified in clause 60 'Compensation events'. Clause 60.1 lists sixteen items as compensation events, including changes to the works information, late possession, failure of the employer to provide materials, failure of the project manager or supervisor to perform their duties as and when required (including replying late to a letter from the contractor).

Under the NEC clause 60.1(12) (clause 12 equivalent in the ICE 6th) in order for the event to qualify as a compensation event the contractor must encounter physical conditions 'which an experienced contractor would have judged to have such a small chance of occurring that it would have been unreasonable for him to have allowed for them'.

The effects of the weather is also a compensation event, giving rise to both time and costs provided the weather is more inclement than might be expected in the previous 10-year period as defined in the schedule of contract data (see Fig. 17.2).

Another new feature of the NEC is the incorporation of the low performance damages, payable by the contractor where the performance of the completed works is lower than specified. The NEC also provides for the payment of bonuses for early completion of sections of the work – again a positive incentive to efficient PM.

---

6. Compensation events • The *weather measurements* are

• rainfall (mm)..

• days with rainfall greater than 5 mm (No.).....................................

• days with minimum air temperature less than 0 degrees Celsius (No.)..

• day with snow lying at......................................hours GMT (No.)........
and these measurements

....

....

....

• The *weather data* are the records of past *weather measurements* which were recorded at ....
and which are available from ....

....

Where no recorded data are available, assumed values for the 10-year return monthly *weather data* are stated here.

• The place where weather is to be recorded is...............................

Figure 17.2   NEC contract data provided by the employer: compensation events – weather measurements. *Source:* New Engineering Contract. Reproduced by permission of the Institution of Civil Engineers

Contractor's design responsibility

The NEC is particularly flexible on design arrangements. The form has been drafted on the assumption that the contractor will always make some design decisions no matter how minor. No extra provisions are required to be made to the standard conditions if the contractor is required to design most or all of the project.

The extent of the contractor's design responsibility can be accommodated in the specification of the work embodied in the works information. This may take the form of a performance specification for a particular component or in the case of a design and build arrangement a statement of performance for the whole project.

Core clauses 20–29 define the contractor's main responsibilities, with clause 21 stating the responsibilities of the contractor with regards to design. The contractor's design of the works is to be in accordance with the requirements of the works information. The contractor's liability to the employer for his design is limited after completion of the whole of the works to the amount stated in the contract data.

Generally the standard of design required from the contractor is a 'fitness for purpose' requirement. If it is anticipated that the contractor will be unable to obtain this type of insurance cover secondary Option M, which is the same as the 'reasonable skill and care' professional indemnity, should be included in the contract.

In order to improve the management of projects the NEC requires that for a firm price contract (measure and value, lump sum) a complete statement of what the contractor will be required to do should be made available to tenderers. Any later instructions given to contractors which clarify gaps or fill in vaguely expressed information should be treated as a variation.

One important aspect missing from the NEC is that no guidance is given on who is responsible for co-ordinating the integration of the contractor's design into the works. The onus would seem to be on the employer's designer to co-ordinate the various elements.

Contractor's programme

The NEC seeks to encourage sound PM by placing a greater emphasis on the contractor's programme. Under the NEC clause 31 the contractor can be required to submit his programme with his tender – this then becomes the 'accepted programme' and is included as a contract document identified in the schedule of contract data. If the alternative procedure is adopted the contractor must submit his programme for acceptance within the period stated in contract data.

The reasons for the project manager rejecting a programme are identified in clause 31.2 which states that the contractor's plans are not practical and do not:

1. Include the information which the contract requires;
2. Represent the contractor's plans realistically; or

3. Show realistic provision for:
    (a) float and other risk allowances;
    (b) health and safety requirements;
    (c) other works information requirements, or
    (d) the procedure set out in the contract.

The programming procedures within the NEC then dictate what must be included within the contractor's programme. The contract states the level of detail required from the contractor: 'for each operation, a method statement which identifies the equipment and other resources which the contractor plans to use'. Failure to meet these provisions is sufficient grounds to reject the programme.

It is noted that NEC clause 50.3 provides that if the contractor does not submit a programme as required under the contract then half the value of work executed to date be retained from the contractor.

Reaction to the NEC

The reaction to the NEC Consultative Document has been encouraging with bodies such as the Yorkshire Water Authority, National Power and the British Airport Authority (BAA) being the first to use the contract in the UK.

After a post-contract audit on a £750 000 project at Southampton airport, the client BAA found that there was a better understanding between the parties, when using the NEC, because of the clear definitions. Significantly this project was completed on time and under budget.

Considerable interest has also been shown by overseas clients particularly in South Africa, Australia and Hong Hong. The Hong Kong Jockey Club, the second largest developer in the colony, has used the NEC on its racecourse redevelopments and is proposing to use the contract on a £100m. plus building project.

Significantly in 1993 an international conference of lawyers, addressed by Martin Barnes, examined the NEC in detail and gave it full marks.

However, some clients and consultants still believe that the NEC has a long way to go before it succeeds in breaking down the long-standing adversarial attitudes in the construction industry. It seems that many parties are adopting a 'wait-and-see' approach before using, or recommending the use of, the NEC. Only time will tell how successful the NEC will be in achieving its stated objective of 'stimulating sound project management'.

Note added in proof: Sir Michael Latham in his report *Constructing the Team*[1] comments that 'the approach of the NEC is extremely effective'. He recommends that the NEC is renamed The New Construction Contract, and that one third of government departments should change to using the new contract by 1998.

Tutorial questions

1. How does the role of the project manager in the NEC compare with that of the engineer in traditional standard forms?
2. In the NEC what is meant by 'compensation events' and how do they differ from the equivalent provisions in the JCT 80 and the ICE 6th?
3. Identify appropriate documentation for a £25m. design and build project for a new sewage works
4. Compare and contrast the provisions for 'risk management' within the NEC with another standard form of contract with which you are familiar
5. The aim of the NEC is to stimulate sound project management; review how the contract aims to achieve this and consider the possible drawbacks to success

Reference

1. Latham M 1994 *Constructing the Team. Final Report of the Government/ Industry Review of Procurement and Contractual Arrangements in the UK Construction Industry* HMSO

Further reading

Allen J D 1991 Civils head latest efforts to curtail contract disputes. *Construction News* 24 January: 6

Barnes M 1991 The New Engineering Contract. *International Construction Law Review* **8**: 247–55

Bill P 1987 Barnes storms another contract. *Building* 18 September: 32–3

Birkby G 1993 New Engineering Contract – not just for engineers. *Project* November: 16–17

Bolton A 1993 Clear for the take off. *New Builder* 2 April: 15

Booth S 1991 The New Engineering Contract, 1991 Cambridge Lecture, *Civil Engineering Surveyor* November: 4–6

Booth S 1993 The 'New' NEC. *Civil Engineering Surveyor* April: 11

Clarke J R 1993 NEC – thoughts and questions. *Civil Engineering Surveyor* July/August: 14

Clarke S J 1993 A study of the mechanics of the New Engineering Contract and its implications for current practice in the construction industry. Dissertation as partial fulfilment of B.Sc. (Hons) Quantity Surveying degree, University of Wolverhampton

Cornes D, and Barnes M 1991 Controversial contract. *New Builder* 31 January: 20–1

Cottam G 1991 ICE set for a user-friendly contract. *Construction News* 24 January: 13

Knowles R 1991 The NEC: a user's guide. *Chartered Quantity Surveyor* June: 8

Lumb K 1993 Legal arguments: the New Engineering Contract. *Building* 23 April: 34

McGowan P H et al 1992 *Allocation and Evaluation of Risk in Construction Contracts* Occasional Paper No. 52, C10B

Mugurian G M 1991 The new civil engineering contract. *Civil Engineering Surveyor* May: 18–20

Nicholson T H N 1992 The New Engineering Contract. *Proceedings of the Institution of Civil Engineers, Civil Engineering* **92**(4), November: 146–8

The New Engineering Contract (Consultative Document), 1991, Thomas Telford

The New Engineering Contract (First Edition), (1993), Thomas Telford

Webster M 1991 New Engineering Contracts, the 1991 Mancunian Lecture. *Civil Engineering Surveyor* October: 12–13

Section H

Case Study

18

Hong Kong Mass Transit Railway

After spending 7 years of one's life on two major projects, which both finished late and cost the client at least double the tendered amount, it is easy to become rather pessimistic about the effectiveness of management control on major construction projects. It came as something of a surprise, therefore, to see the complete contrast on the Hong Kong Mass Transit Railway (HKMTR) project and to discover that not only major but mega projects could be successfully managed, controlled and completed on time and within budget.

Brief history

The need for a mass transit railway to fulfil Hong Kong's future transport requirements was first identified in the Hong Kong government's Mass Transport Study in 1968.

In the early 1970s a major Japanese trading company offered to design and construct the whole of the Initial System under a single fixed-price turnkey arrangement. However, this approach was aborted following the withdrawal of the turnkey contractor in the wake of the 1973 Middle East oil crisis. This led the corporation (HKMTRC) to utilise a more conventional contract strategy with the project being split into stages of 3–4 years' duration with each stage subdivided into sections comprising six to eight major contracts.

Stages 1 to 3 of the HKMTR, comprising the Modified Initial System, the Tsuen Wan extension and the Island Line, were constructed during the period 1975–84 and comprised a total of 37 stations (mainly underground), with high-rise property developments above the stations, massive site formations (e.g. 5 million m³ of rock excavation at Kornhill), large sea reclamations, an immersed tube harbour crossing and elevated roads.

Since 1984 the second immersed tube – the Eastern Harbour crossing – with an additional station have completed the system loop from the Kowloon mainland to the Hong Kong island. This stage, which was completed in 1989, was constructed using a build–operate and transfer

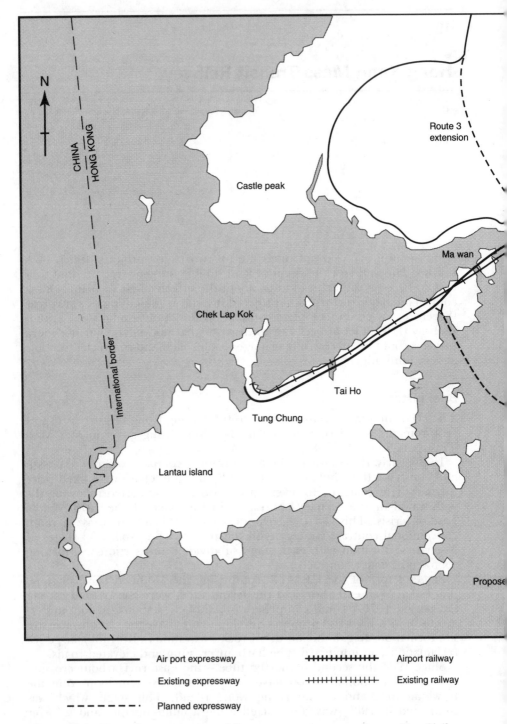

Figure 18.1   Hong Kong, present and future transport routes. *Source: New Civil Engineer*, 25 June 1992, p. 22. Reproduced by permission of Thomas Telford

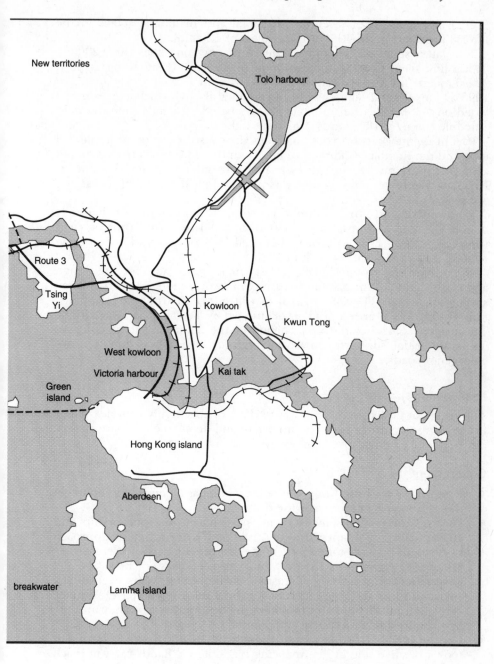

New territories

Tolo harbour

Route 3

Tsing Yi

Kowloon

Kwun Tong

West kowloon

Victoria harbour

Kai tak

Green island

Hong Kong island

Aberdeen

breakwater

Lamma island

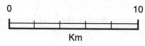

0   10

Km

(BOT) arrangement. The cumulative total cost of all the stages in the HKMTR project up to 1991 was in excess of £2000m.

The latest stage of the railway development comprises the £2000m. 34 km link from Central and Kowloon to the new Hong Kong airport. This Lantau and Airport Railway project, with the first contracts invited in 1993, is consultant designed with conventional B of Q which are prepared based on CESMM3. Interim payments are based on a milestone payment schedule completed by contractor in his bid.

Part of the new airport route includes the construction of an immersed tube and the Rambler Channel bridge. Both these projects are design and construct fixed price lump sum contracts not subject to remeasurement. Payment is at fixed monthly intervals provided predetermined milestones are met.

The Lantau and Airport Railway project when complete will nearly double the size of the existing system and will feature 19 major civils contracts with 30 E&M contracts managed by a project and engineering team of 1200 people at its peak in 1995.

In addition the Hong Kong government is responsible for the design and construction of the Lantau Crossing, the world's longest road and rail suspension bridge. The £550m. project, which forms part of the HKMTRC railway link, was awarded to a consortium led by Trafalgar House in 1992. The project comprises the construction of the 2 km long double-decker Tsing Ma bridge supporting a six-line highway, two railway lines and two emergency roads.

Reasons for success

The reasons that led to the success, to date, of this highly complex major urban infrastructure project are important and demand examination; they can be summarised as given under the headings below.

Government

- Uniqueness of Hong Kong where everyone has a determination to get things done no matter what the difficulties.
- Positive government involvement, particularly on sponsorship and guaranteeing of finance. In contrast the UK government approach is typified by the comments made by Sir Richard Luce, former Minister of Arts on the British Library project disaster: 'But above all, governments through the Treasury, have over the years been totally reluctant to see the long-term advantage of releasing taxpayers' money early and quickly in order to obtain the best value for money and the best chance for a successful completion'.[1]
- On the HKMTR project even though the Hong Kong government took a positive role in the financing it adopted a 'hands-off' approach with the client's PM team left to manage the project.
- Stable economic and political climate (pre-1985). With the approach of 1997 and the Chinese take-over of Hong Kong the construction of the new airport has become one of the key items on the agenda which is reflected in the current uncertainty surrounding the project.

Finance

- Clear objectives on the financial viability of the project. The Corporation is financially sound. In the 1991 Annual Report[2] it was able to record that it was one of the very few underground mass transit systems in the world which generated unsubsidised fare revenue covering the operating and depreciation costs with a satisfactory operating profit. In 1993 the railway was carrying in excess of 2 million passengers each day during its 19 hours' operating period.
- Property development also made a significant contribution to the funding of the railway system. The 1991 Annual Report recorded that all the property developments on the existing railway were completed in 1990 with a cumulative final profit margin amounting to £340m.
- Provisions were made in the contracts for contractor finance – utilising the export credit arrangements offered by foreign governments.
- All prospective tenderers required to submit both technical and financial data to enable them to be prequalified for specific contracts. Guarantees required for joint ventures and from parent companies.
- On the later stages of the project all contracts were let based on fixed prices expressed in Hong Kong dollars, thus limiting the corporation's exposure to fluctuating currency rates.
- Contractors' tenders compared on a net present value (NPV) basis. Using the interim payment schedule method (see fuller description under 'Contract') the corporation was able to identify the 'best value tender' based on each contractor's required cash flow during the construction period; this was not necessarily the lowest tender bid.

Organisation and method

- The HKMTRC decided from the outset that retention of control of all contracts and the exercise of strict financial and contractual disciplines were crucial to achievement of completing the project on time and within budget.
- the MTRC retained full control of the construction by setting up a strong PM structure to control the project. This was achieved through four groups:
  Engineering group: responsible for all the pre-contract activities of design – both civils and architecture;
  Project group: responsible for managing the project;
  Contracts group: responsible for tender documentation and assessment and prior to the award as the MTRC's contractual adviser;
  Programming group: responsible for overall strategic and contract programming.
- The project group organisation was slightly modified in the light of experience; on the third stage of the project, the Island Line, the management structure was as follows:
  The engineering and project director was a member of the corporation's executive board.
  The engineering manager was appointed the 'engineer' in addition

to retaining management responsibility for all civil design and co-ordination with the E&M function.

Two civils project managers, reporting directly to the engineering and project director shared the responsibility for the civils construction and one E&M project manager was responsible for overseeing the design and construction of the E&M works.

● Site supervision controlled by five civils construction managers (one on each section) and one E&M construction manager. All the powers of the engineer were delegated to the construction managers except those relating to final completion, maintenance and the resolution of disputes.

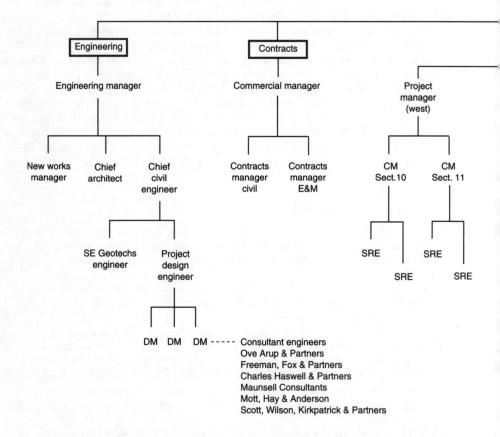

CM Construction manager   SRE Senior resident engineer   SE Senior engineer   DM Design manager

Relative positions do not indicate seniority

Figure 18.2   Hong Kong Mass Transit Railway, Stage 3, Island line, organisational structure. *Source*: Donald et al.[3] Reproduced by permission of the Hong Kong Mass Transit Railway

- The engineer's representative functions were exercised by the senior resident engineers.
- A typical post-contract site-based PM section comprised: a construction manager, two senior resident engineers, resident engineers and assistants, construction planner, land surveyors, inspectors, safety officer, contract administrator and quantity surveying staff.
- Project split into 3–4-year-long stages, each stage split into sections, each section managed six to eight contracts valued at £250m. total (at 1982 prices).
- Number and size of contracts encouraged the participation of overseas

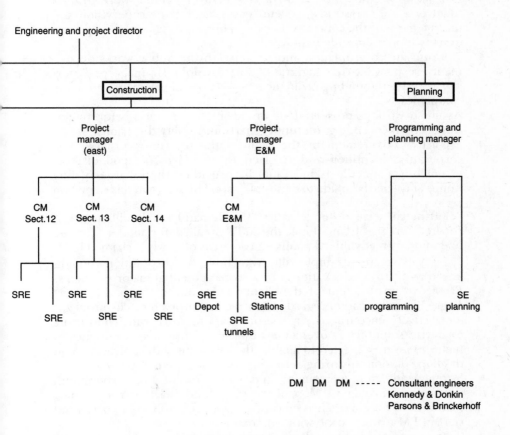

contractors in difficult areas enabling local contractors to execute the more straightforward work.

- Element of competition between Japanese, British, French, Australian and local contractors and even between different site agents on the same projects. Constant director involvement in projects.

- E&M content of the project dealt with on a functional basis with each E&M contractor responsible for a particular element over the full length of the project stage, e.g. overhead current collection system, electrical power system. E&M contracts let on basis of reference designs and performance specifications.

- Detailed procedure notes and a cost control manual were drafted to cover all administrative systems necessary for the supervision and management of the contracts. Good-quality accurate reports to project executive on a monthly basis.

- Appropriate co-ordination and liaison with the public works departments responsible for highways, water supply, drainage and port works and with building ordinances, police, fire department and utility companies.

- As much work as possible done on advanced contracts before major works commenced, e.g. on tunnel shafts and utility diversions.

- Pro-active involvement by the client at the pre-contract stage. The corporation identified and resolved the problem of spoil disposal (3 million m³ on the Island Line) by providing three barge-loading ramps at the harbourside to deposit the spoil in designated reclamation areas.

  Furthermore in order to satisfy the demand for 1 million m³ of concrete on the Island Line, the MTRC called for tenders for the operation of ready mixed plants on two areas of newly reclaimed land.

  The above two examples illustrate how the corporation thought ahead and provided solutions to overcome potential major problems.

- Pro-active involvement of the client's PM team at the post-contract stage. The corporation regarded contractors' problems as its problems and actively encouraged supervisory staff to work with contractors to devise solutions to overcome obstacles to progress. Construction managers were expected to analyse the problems with contractors and develop solutions ensuring recovery of any lost time.

- Strong leadership, personality and know-how of the corporation's construction managers. Most of the corporation's senior project staff had worked for contractors on similar projects. Sufficient experienced staff in PM team to cope with workload.

- A comment of Professor Peter Thompson of UMIST is particularly appropriate: 'Experience suggests that selecting project staff with a positive, forward looking attitude, coupled with commercial awareness and experience of similar projects, is more important than for them to have detailed technical knowledge.'[4]

- As a large volume of equipment was secured from the UK an HKMTRC office was established in the UK. This office was headed by a contracts manager (Europe) and was responsible for inspection and progress chasing.

Communications

- Timely decision making with *ad hoc* as well as regular meetings. Typical were the twice-weekly early-morning communication meetings between the divisional director and all the senior managers responsible for design, site supervision, programming and commercial matters.

    These regular meetings ensured that all the senior managers were aware of the current situation on the project. An element of competition was also introduced between senior managers – no one wishing to be responsible for any delay to the project.
- A further example of sound lines of communication included the quarterly meetings at which all the E&M contractors and the track-laying contractor were represented at director, and frequently at managing director, level. This forum enabled outstanding problems of concern, and matters affecting interfaces, to be discussed and the requisite corrective action to be initiated.
- 'Indeed the essence of successful construction management can be identified as the early identification of problems and the rapid elevation of them, to whatever levels were required for solution, be it at Resident Engineer, Construction Manager or Directorate level', Donald, W R. et al.[3]
- Good levels of communication at all levels were further typified by the regular communication meetings between the project director and the PM team on each section – after working hours with light refreshments provided. These informal gatherings encouraged team building and helped to reinforce the corporate identity.
- Regular contacts with the public through public relations department.

Technology and design

- Generally the use of traditional and proven technology.
- Corporation's open attitude to contractors' alternative design proposals, particularly if this could secure a time saving without a reduction in quality, e.g. the use of the new Austrian tunnelling method (NATM) involving sprayed concrete in lieu of concrete linings in certain rock tunnels.
- Comprehensive soil investigation in order to minimise design changes and unforeseen ground condition claims.
- Due to vast quantity of approvals required from the building authority and the time constraint the corporation provided a suggested temporary works design in the tender documents utilising knowledge obtained from the earlier contracts. These temporary works were measured and included in the B of Q (fundamental principle of contract strategy that the party best able to assess a risk should carry it).
- Project designed by consultants, design complete and frozen. However, some changes were inevitable, e.g. for design developments, interfaces with E&M work, alterations in operational requirements or unforeseen conditions.
- Management of consulting engineers' design performed by design managers with small teams of engineers attached to each section on

site – critical on stage two where construction commenced before the design was complete.

- Strict control procedure required proposer to submit justification for change complete with estimate of time and cost. Proposal considered at twice-weekly divisional review meetings allowing principle to be considered at the highest level before expenditure incurred.

Contract

- The major contracts for the underground works for the first phase of the project were let utilising contractor's design and build based on the engineer's general arrangement drawings.
- The design and build approach was adopted in order to secure the earliest possible start and completion dates and to allow economic use of contractors' preferred methods of construction and available construction equipment. Lump sum prices were broken down against major activities with a brief schedule of rates for valuing extra works; however, in practice the scheduled items did not always reflect the varied work.
- On the latter stages of the project the civils contracts were let on the basis of full working drawings being prepared by the engineer (in practice the consulting engineers appointed by the employer). Variations were thus kept to an absolute minimum.

  Tenders were based on B of Q which were measured in accordance with the MTR method of measurement (similar to CESMM2 with modifications to the tunnelling section and the addition of an architectural works section).

- Contract based on known standard conditions of contract: FIDIC with some amendments to include ICE 5th and the Hong Kong Public Works Department conditions.
- Careful vetting of main contractors' subcontractors by the corporation's contracts section prior to their appointment. This is in contrast to the provision of the ICE 6th (clause 4 (2)) which only requires the

---

Activity Number: 236/021/6

Activity Title: New STN. Station Foundations

General Description

Reinforced concrete pile caps of various plan dimensions and depths for station columns and walls. Reinforced concrete ground beams between pile caps. Approximate concrete volume 1620 m³.

Lump sum (carried to summary) HK$_____

Contents of	Labour	_____ %
lump sum	Materials	_____ %
	Other coats	_____ %

Figure 18.3   Example of typical activity schedule item

engineer to be notified in writing prior to the subcontractor's entry on to the site.

- A monthly admeasurement payment system was not used, interim payments were linked to progress. The activities under each contract were segregated into major cost centres (between four and six), including one dealing with mobilisation/preliminary items and provisional sums.
- Each contractor was required with his bid to submit an 'interim payment schedule' (IPS) detailing the percentage of each cost centre value he wished to receive during successive months of the contract. Subject to satisfactory progress these proportions of the cost centre value, adjusted as necessary to deal with variations, were paid automatically each month. A parallel admeasurement process allowed appropriate adjustment as activities were completed.
- If specified milestones were not met, all payment on that cost centre ceased until the milestone event was achieved. There was therefore compelling financial reason for adherence to the required programme.

Month	Cost centre A	Cost centre B	Cost centre C	Cost centre D
1	1.00	1.00	2.00	0.00
2	2.00	3.00	5.00	0.00
3	3.50	5.00	10.00	0.00
4	5.00	9.00	15.00	1.00
5	7.00	12.50	20.00	3.00
6	9.00	17.00	25.00	5.00
20	47.00	90.00	93.00	32.00
21	49.00	95.00	100.00	35.00
22	51.00	100.00		37.00
34	98.00			90.00
35	99.00			95.00
36	100.00			100.00

Figure 18.4   Extract from typical completed interim payment schedule

- The use of the interim payment schedule system enabled the corporation to compare tenders using the NPV method; furthermore the corporation knew its potential cash flow whilst the contractor was assured of his cash flow subject to performance. For full details of the IPS system see the author's article.[5]
- A general policy to avoid arbitration or litigation if at all possible. Early settlements of final accounts – often upon contract completion. Good claims management and reporting system.
- In exceptional circumstances 'commercial settlements' were made prior to completion, direct between the client and the contractor. Full payment of the 'commercial settlement' often linked to the satisfactory progress of the outstanding work and the achievement of critical dates.

Construction programme

- Opening date fixed and unalterable.
- Strategic programme for whole stage of project identified everything critical in the early stages.
- Construction programming identified key dates, e.g. contract interfaces and interfaces with government and utility companies' works.
- Project programmes identified relevant 'critical' and 'milestone dates'. Key dates included milestone dates in individual contracts with payments suspended if the required progress was not achieved. Key dates also included critical dates applicable to a stage of completion and contract interfaces for which liquidated damages would apply. Typically a major civils contract would contain more than 10 critical dates and over 60 milestone dates.
- Contractor required to submit comprehensive contract programme in network form within 60 days of contract award. Contractor required to update programme periodically to demonstrate how he intended to fulfil his obligations.
- Contractor's programmes checked for logic and productivity forecasts. If unacceptable contractor notified and asked to resubmit (as the new provision within clause 14 of the ICE 6th).
- Progress monitored by regular weekly and monthly meetings. Internal reporting system ensured corporation kept aware at regular monthly meetings of all time or cost overruns.

Cost control

- Success of system depends on control of costs, not merely monitoring.
- System involved establishing budget based on accurate estimates, preparing accurate reports of actual expenditure throughout the project, control of progress of the works, control of expenditure on variations, control of unavoidable additional costs.
- The corporation's cost control manual required that matters giving rise to possible cost changes were identified as early as possible, allowing cost–benefit analysis to be undertaken. The corporation's executive

were required to give approval before any significant additional expenditure was instructed.

- Accurate monthly reports on each contract gave the updated estimated final contract prices, including for all admeasurement, approved cost changes and expenditure against provisional items.
- System allowed considered judgement on acceptability of proposed variation or acceleration proposal to be considered within the framework of the financial state of the whole project.

Conclusions

It can be stated with some certainty that the design, planning and construction of the Hong Kong Mass Transit Railway have been one of the great success stories in international construction PM in the last three decades. Each stage was completed in, or before, the scheduled time and was completed within budget. Furthermore the client was highly satisfied with the finished project which met its business objectives.

We can do no better than conclude by summarising the reasons for the success of the project as identified by the HKMTRC's Chief Civil Engineer Mr D. J. Sharpe:[6]

- clearly defined objectives and targets
- simple yet rigid corporate values of 'to standard, on time and within budget'
- adequately resourced organisation and structure
- regular and meaningful communication at the appropriate levels
- fast and positive action when things go wrong or when time and/or money is at risk
- linking of progress to payments via milestones
- use of proven technology
- belief in the necessity of quality assurance
- government's positive support of MTRC as an autonomous body, with minimum intervention
- teamwork

How appropriate that the last word in this book should be 'teamwork'.

References

1. Luce R 1992 Library's lessons for the future. Letters to the Editor *The Daily Telegraph* 4 December
2. Mass Transit Railway Corporation of Hong Kong 1992 *Annual Report 1991*
3. Donald W R, Hillier A M, Greig W J Project and construction management on Hong Kong Mass Transit Railway (unpublished paper)
4. Thompson P 1991 The client role in project management. *International Journal of Project Management* 9(2), May: 90–2
5. Potts K F 1988 An alternative payment system for major 'fast track' construction projects. *Construction Management and Economics* 6: 25–33

6. Sharpe D J 1987 Urban railways and the civil engineer, Session 3, Construction, Paper 10: Completing on time – construction management. *Proceedings of Conference, Institution of Civil Engineers*, London, 30 September–2 October 1987

Further reading

Edwards J T et al 1980 Hong Kong Mass Transit Railway, Modified Initial System: system planning and multi contract procedures. *Proceedings, Institution of Civil Engineers* Part 1, **68** (November): 571–98

Edwards J T et al 1982 Discussion, Hong Kong Mass Transit Railway, Modified Initial System. *Proceedings, Institution of Civil Engineers* Part 1, February: 87–98

List of cases

Listed in the same order as in the book:

Rosehaugh Stanhope (Broadgate Phase 6) plc and Rosehaugh Stanhope (Broadgate Phase 7) plc v. Redpath Dorman Long Ltd (1990) 50 BLR 69
Simplex Concrete Piles Ltd v. St Pancras Borough Council (1958) 14 BLR 80
Dudley Corporation v. Parsons and Morrin Ltd (1959) BIN Feb 17 1967
Farr (AE) Ltd v. Ministry of Transport (1965) 5 BLR 94
Mitsui Construction Co Ltd v. The Attorney General of Hong Kong (1986) 33 BLR 1
English Industrial Estates Corporation v. Kier Construction Ltd (1991) 56 BLR 93
Robinson v. Harman (1848)
Hadley v. Baxendale (1854) 9 Exch 341
The London Borough of Merton v. Stanley Hugh Leach (1985) 32 BLR 51
Pearce (C J) & Co Ltd v. Hereford Corporation (1968) 66 LGR 647
Tate & Lyle Food Distribution Ltd v. GLC (1982) 1 WLR 149
Minter (F G) Ltd v. Welsh Health Technical Services Organisation (1980) 13 BLR 1
Rees and Kirby Ltd v. Swansea City Council (1985) 30 BLR 1
Crosby (J) & Sons Ltd v. Portland Urban District Council (1967) 5 BLR 121
Mid Glamorgan County Council v. J. Devonald Williams and Partners (1992) CILL 722
Bacal Construction (Midlands) Ltd v. Northampton Development Corporation 8 BLR 88
Turriff Ltd v. Welsh National Water Development Authority (1980) (unrep)
Yorkshire Water Authority v. Sir Alfred McAlpine & Son (Northern) Ltd (1985) 32 BLR 114
Glenlion Construction Ltd v. The Guinness Trust (1987) 39 BLR 89
BLR   Building Law Reports
BIN   Building Industry News
CILL   Construction Industry News Letter
Exch   Exchequer Reports
LGR   Local Government Reports
WLR   Weekly Law Reports

Further reading

Abrahamson M 1979 *Engineering Law and the ICE Conditions* 4th edn
  Elsevier Applied Science Publishers
Bingham A et al Legal columns in *Building* (weekly)
Duncan Wallace I N 1970 *Hudson's Building and Engineering Contracts* 10th
  edn Sweet & Maxwell
Duncan Wallace I N 1979 *Hudson's Building and Engineering Contracts* 10th
  edn, First Supplement, Sweet & Maxwell
Eggleston B 1993 *The ICE Conditions of Contract: Sixth Edition, A User's Guide*
  Blackwell Scientific Publications
Knowles R 1992 *Thirty Crucial Contractual Issues and their Solutions* Knowles
  publications
Knowles R Legal column in *Chartered Quantity Surveyor* (monthly)
May A (ed) 1992 *Keating on Building Contracts* 5th edn Sweet & Maxwell
Powell-Smith V Law column in *Contract Journal* (weekly)

Useful addresses

Association of Cost Engineers
  Lea House
  5 Middlewich Road
  Sandbach
  Cheshire CW11 9XL
  Tel: 0270 764798
  Fax: 0270 766180

Association of Project Managers
  85 Oxford Road
  High Wycombe
  Bucks HP11 2X
  Tel: 0494 440090
  Fax: 0494 528937

Building Bookshop
  26 Store Street
  (off Tottenham Court Road)
  London WC1E 7BT
  Tel: 071 637 3151
  Fax: 071 636 3628

Building Cost Information Service
  85/87 Clarence Street
  Kingston upon Thames
  Surrey KT1 1RB
  Tel: 081 546 7554
  Fax: 081 547 1238

Building Employers Confederation
  82 New Cavendish Street
  London W1M 8AD
  Tel: 071 580 5588
  Fax: 071 631 3872

Chartered Institute of Arbitrators
  International Arbitration Centre
  24 Angel Gate
  City Road
  London EC1V 2RS
  Tel: 071 837 4483
  Fax: 071 837 4185

Chartered Institute of Building
  Englemere
  Kings Ride
  Ascot
  Berkshire SL5 8BJ
  Tel: 0344 874640
  Fax: 0344 23467

Construction Industry Computing
Association
  Guildhall Place
  Cambridge CB2 3QQ
  Tel: 0223 311246
  Fax: 0223 62865

CTA Services (Swindon) Ltd
  Belmont House
  11/12 Devizes Road
  Old Town
  Swindon, Wilts SN1 4BH
  Tel: 0793 610506
  Fax: 0793 481281

European Construction Institute
  Loughborough University of
  Technology
  Sir Arnold Hall Building
  Loughborough
  Leicestershire LE11 3TU
  Tel: 0509 222620
  Fax: 0509 260118

Major Projects Association
 Templeton College
 Kennington
 Oxford OX1 5NY
 Tel: 0865 735422
 Fax: 0865 326068

Federation of Civil Engineering
Contractors
 Cowdray House
 6 Portugal Street
 London WC2A 2HH
 Tel: 071 404 4020
 Fax: 071 242 0256

Institution of Civil Engineering
Surveyors
 26 Market Street
 Altrincham
 Cheshire WA14 1PF
 Tel: 061 928 8074
 Fax: 061 941 6134

Institution of Civil Engineers
 1–7 Great George Street
 Westminster
 London SW1P 3AA
 Tel: 071 222 7722
 Fax: 071 222 7500

Royal Institute of British Architects
 66 Portland Place
 London W1N 4AD
 Tel: 071 580 5533
 Fax: 071 255 1541

Royal Institution of Chartered
Surveyors
 12 Great George Street
 London SW1P 3AD
 Tel: 071 222 7000
 Fax: 071 222 9430

UMIST
 Continuing Education Office
 PO Box 88
 Manchester M60 1QD
 Tel: 061 200 3995
 Fax: 061 228 7040

University of Wolverhampton
 School of Construction,
 Engineering and Technology
 Wulfruna Street
 Wolverhampton WV1 1SB
 Tel: 0902 321000
 Fax: 0902 322680

Index